U0782179

低碳行动

绿色美食

就在你身边

李玲玲 著

Cecilia Lingling Li

Low-Carbon Living:
Green Cuisine
at Your Fingertips

SPM
南方传媒
广东科技出版社
全国优秀出版社
· 广 州 ·

图书在版编目（CIP）数据

低碳行动：绿色美食就在你身边 / 李玲玲著.
广州 ： 广东科技出版社，2025. 4. -- ISBN 978-
7-5359-8442-5

Ⅰ．TS971.202.65

中国国家版本馆 CIP 数据核字第 2025F5Y989 号

低碳行动：绿色美食就在你身边

Ditan Xingdong: Lüse Meishi Jiu Zai Ni Shenbian

出 版 人：严奉强

项目统筹：颜展敏

责任编辑：彭秀清

装帧设计：友间文化

插 画：李玲玲（Cecilia Lingling Li） 尹 然 宇玲文化

责任校对：韦 玮

责任印制：彭海波 林记松

出版发行：广东科技出版社

（广州市环市东路水荫路11号 邮政编码：510075）

销售热线：020-37607413

https://www.gdstp.com.cn

E-mail：gdkjbw@nfcb.com.cn

经 销：广东新华发行集团股份有限公司

印 刷：广州市岭美文化科技有限公司

（广州市荔湾区花地大道南海南工商贸易区A幢 邮政编码：510385）

规 格：787 mm×1 092 mm 1/16 印张15.5 字数270千

版 次：2025年4月第1版

2025年4月第1次印刷

定 价：88.00元

如发现因印装质量问题影响阅读，请与广东科技出版社印制室联系调换（电话：020-37607272）。

序言

　　我们常说，饮食即文化。一日三餐，一蔬一饭，看似寻常，却浓缩着一个地方、一个民族对自然的理解与生活的智慧。

　　而今，我们正站在生活方式转型的十字路口。气候变暖和资源紧张，让"低碳生活"从理念变为行动；曾经的那种"高消费＝好生活"的观念，正被越来越多的人以"有温度、有节制、有质感"的方式悄悄改变——从"吃得多"转向"吃得好"，从"全球买"回归"本地食"；大疫之后，我们也重新拾起对厨房的珍惜，对健康食材、安心烹饪、陪伴家人、共享一餐的美好渴望。此时此刻，我们比任何时候都更需要一种温和而坚定的声音，引导我们重新思考：如何吃饭，如何生活。

　　《低碳行动：绿色美食就在你身边》正是在这样的背景下诞生的。它不是一本单纯讲菜谱的工具书，而是一份兼具知识、故事与温度的生活提案，像一阵春风，轻轻吹进厨房，也悄悄唤醒我们与日常的连接。

　　本书以广东为起点，放眼全球，从岭南饮食中节能低碳的智慧讲起，再到日本、法国、以色列、古巴、秘鲁、希腊、意大利等地的绿色实践与烹饪趣事——从广东主妇口中的"知悭识俭"，到"不时不食"的菌菇宴；从剩料变甜品的法式巧思，到好玩又促进亲子关系的意大利鸟基……作者细致梳理了这些在地饮食文化中可借鉴的环保经验。她以贴近生活的笔触告诉我们：绿色低碳不等于清汤寡水，不是只吃青菜不沾荤腥，"适度"才是硬道理！真正的低碳饮食，是对土地、对资源、对味蕾的体贴与尊重。

玲玲的文字，有电视人的镜头感，也有生活家的温润与亲切。她会说，广东人炒菜心加一点猪油渣，是为了让素菜也有"荤香"；她会讲，以色列人制作沙克舒卡，只需将前一天剩下的各种蔬菜边角料，如青椒、黄椒、小红辣椒、香菜、樱桃番茄以及蒜片等炒在一起，再加上两个荷包蛋就是一道美味。她没有说教，只是轻轻地推你一把，在你准备食材、翻动锅铲、对待剩饭的时候，悄悄完成一次绿色的选择。

　　认识玲玲已经十余年。她是一位资深媒体人，也是一位优秀的"斜杠青年"：纪录片制作人、中英双语主持人、作家、文化交流者。多年来，她走访三十多个国家，深入厨房，走进菜园，与驻粤总领事、米其林主厨、本地家庭主妇促膝交谈，探寻一蔬一饭背后的生活智慧。在书中，她巧妙地让岭南的饮食传统与世界各地的低碳理念交相辉映，既有田野调查的奔忙和欣喜，也有生活体悟的温度和快乐。

　　本书的可贵之处还在于，它不仅是一本书，更像是一位会做饭、懂生活、爱地球的朋友。读这本书，像是打开一扇窗，窗外是另一种可能的日常——它不极端、不教条，而是悄悄告诉你："你其实可以这样生活。"而这种生活，并不需要剧烈改变，不必放弃味蕾的愉悦，也无须追逐昂贵食材。哪怕只是一杯自制的生姜柠檬水，一碗不浪费食材的广东炖汤，也都是对地球母亲的一份温柔回应。

　　值得一提的是，这是一本以中英双语呈现的低碳饮食文化书籍，内容原汁原味，表达流畅自然。书中还配有专为章节内容量身创作的精美插画——碧绿菜心跃然纸上，冬阴功汤香味飘散，色彩温润、画面生动，令人一翻书页便"眼睛先吃一口"。部分插画展现了作者在广州的成长片段，也映射出岭南家庭文化的底蕴。它让知识有趣，视觉有味，也让阅读本身成为一种轻盈的享受。

作为出版人，我由衷敬佩玲玲对内容的打磨与用心。她不是在喊一个环保口号，而是用脚踏实地的生活实践，将"低碳"从标语里轻轻取出，安放进每一位读者的饭碗里。

衷心希望更多读者能在这本书中，找到属于自己的节奏与选择。正如作者所说："从餐桌开始改变，生活就会跟着变好一点。"你不必一下子走得很远，不妨从今晚的一顿饭开始。

严奉强

广东科技出版社社长

Preface

We often say that food reflects culture. In our three daily meals—simple as they may seem—lie the quiet expressions of how a people relate to nature and to life itself.

Today, as the world reimagines how we live, low-carbon living is moving beyond a concept—it is becoming a conscious way of life. The old belief that "more consumption means better living" is giving way to a quieter wisdom: to live is to live thoughtfully. That means eating better, not more; choosing what's local and in season, not just what's available from afar. In the wake of the pandemic, we've rediscovered the comfort of home kitchens, the honesty of real ingredients, and the deep joy of gathering around the table.

It's under this renewed awareness that *Low-Carbon Living*: *Green Cuisine at Your Fingertips* was born. This is more than a cookbook but rather a thoughtful companion—part guide, part collection of lived stories infused with warmth. It steps into the kitchen like a spring breeze, gently rekindling our relationship with food and the everyday life.

The journey begins in Guangdong and reaches kitchens around the world—starting with the low-carbon living wisdom found in the cuisine of Lingnan to Japan, France, Israel, Cuba, Peru, Greece, Italy,etc.—sharing how different cultures bring low-carbon wisdom to life. For example, there are stories of frugal Cantonese housewives transforming fruit scraps into decorations, soup or tea, seasonal mushroom dishes, French desserts made from stale bread, and playful gnocchi made with children. These pages offer practical, graceful ways to eat thoughtfully—nourishing the planet, our bodies, and the joy of flavor.

Lingling writes with the eye of a filmmaker and the heart of a home cook. Whether she's capturing the rich aroma of lard dregs in stir-fried Choy Sum or describing how Israelis transform leftover vegetables into Shakshuka, her tone is warm, never instructive. She gently encourages us to make better, greener choices—whether it's in selecting ingredients, stirring during cooking, or using up leftovers.

I've known Lingling for over a decade. She is a seasoned media professional and a true cultural storyteller—a documentary producer, bilingual TV host, and writer with a rare gift for connection. Having traveled to more than thirty countries, she has cooked with Michelin-starred chefs, spoken with people from all walks of life, diplomats,and shared meals with everyday families. In this book, she brings together the wisdom of Cantonese cuisine and traditions, and the global language of sustainability, grounding it all in lived experience.

This is not just a book—it's a friend who cooks with care, lives with purpose, and tends to the Earth. It doesn't preach or push. It simply says: there is another way to live—and it can begin with something as simple as a meal. You don't have to go far all at once—even something as simple as a cup of homemade lemon ginger drink or a bowl of Cantonese soup that puts no ingredients to waste, is a gentle response to Mother Earth.

With bilingual text and vibrant cartoon illustrations, each chapter offers something to savor, both visually and emotionally. From bright green Choy Sum to the spice and steam of Tom Yum Goong, the book invites the reader's eyes and palate to journey together. Some of these illustrations capture scenes from the author's childhood in Guangzhou and reflect the rich heritage of Cantonese family culture.

As someone who has been in publishing for a long time, I deeply admire Lingling's precision and warmth. This isn't a book of empty slogans. It brings the idea of "low-carbon"down to the plate— where it can nourish, inspire, and gently shift our habits.

I hope you find in these pages your own rhythm and something that suits you. As the author once mentions: "Change begins at the table and life begins to taste better." Perhaps, we can all begin this very evening, with a single and thoughtful dish.

Yan Fengqiang
President of Guangdong Science and Technology Press

目录
CONTENTS

第 3 章　低碳取材
Chapter 3　Sourcing Ingredients with Smaller Carbon Footprints

4

第 4 章　这些佳肴烹饪起来节能、省时又省心
Chapter 4　Gourmet Dishes Which are Energy-Efficient, Time-Saving, and Worry-Free

第 5 章　美食中的废料利用
Chapter 5 Enjoying Delicious Food While Minimizing Food Waste

我小时候在广州的家中与父母吃饭

Here's an illustration of me when I was a kid. I am sharing a meal with my parents in our little place in Guangzhou

第 1 章
Chapter 1

广东饮食文化中的低碳与环保

Green and Low-Carbon Living in Cantonese Food Culture

　　广东，物产丰富，气候宜人。当地人的饮食习惯开放包容，除了喜爱本地食材，也乐于接受和尝试来自其他省市乃至全球各地的美食。因此，行走在广东的大街小巷，你几乎可以吃得到各地的风味佳肴。如果喜欢，广东人每天每餐都能品味不同的美食，一周七天下来不重样。然而，在广东人的餐桌文化中，有两样食物是不可或缺的：青菜和白肉，比方说，鸡肉。任何来自鸡的部位都能成为老广餐桌上的美味佳肴。此外，广东人对鱼类也情有独钟，这包括了塘鲜、河鲜和海鲜。

　　联合国粮食及农业组织的数据显示，红肉生产过程中排放的甲烷是一种主要的温室气体。与红肉及奶制品生产相比，蔬菜种植的平均碳排放量要低得多，而牛肉的碳排放密度尤其高。专家指出，牛羊养殖对气候变暖的影响大约是蔬菜种植、鸡肉或鱼类养殖的十倍。因此，无论是有意为之还是自然而然，广东人喜食青菜、鸡肉和鱼类的饮食习惯，实际上在减少温室气体排放方面发挥了积极的作用。

　　广东人不仅在食材选择上注重绿色与健康，还偏爱节能低排的烹饪方式，像清蒸、白灼、清炒等方法，耗时短、能量消耗低，同时最大程度地保留了食材的鲜美本味。此外，广东人热衷于在家种植蔬菜，善于物尽其用（如晒制陈皮煲汤等），这些环保、低碳的生活方式无疑为建设现代绿色社会提供了借鉴。

　　本章将从食材的选择、获取方式、烹饪方法以及二次利用等四个方面向读者概括地介绍广东本地饮食文化中值得借鉴的绿色与低碳理念。

Guangdong is rich in agricultural products and blessed with a mild climate. Guangdong people in general aren't picky when it comes to food. They have quite open and inclusive taste buds. In Guangdong, foods from other provinces, cities, countries and regions can be easily found. If they like, Guangdong people can eat something different in every meal 7 days a week. On the other hand, in Cantonese food culture, two things are almost indispensable and appear a lot on the meal table: one is green vegetables, and the other is white protein/meat, such as chicken. Any part of the chicken can be made into a delicious dish on a Cantonese table. Guangdong people also eat a lot of fish, freshwater products, and seafood.

According to the data provided by the Food and Agriculture Organization of the United Nations, methane emitted from red meat production is an important greenhouse gas. Compared with the carbon emission intensity of red meat and dairy production, the average carbon emission of vegetable planting is much lower. Among red meat, beef production has the highest carbon emissions. Most experts believe that the contribution of cattle and sheep breeding to climate warming is more than 10 times higher than that of vegetable planting, chicken or fish breeding. Whether intentionally or unintentionally, the habit of eating vegetables, chicken, freshwater products and seafood amongst the Cantonese has helped reduce greenhouse gas emissions contributing to environmental protection.

In addition to eating more green and healthy foods in daily living, Cantonese also prefer to use energy-efficient and low emission cooking methods to prepare various kinds of foods. The usual ones include steaming, blanching and quick stir-frying. These techniques consume less energy and take little time to do. In a blink of an eye, the dishes are cooked and the natural tastes of the ingredients are preserved well. Moreover, the local people's practice of growing vegetables at home and maximizing the use of food waste like sun-drying the mandarin peel for soup cooking are other low-carbon living and environment-friendly lifestyles that we can learn from.

This chapter will provide an overview of the green and low-carbon food culture in Guangdong from four aspects: food ingredient choosing, food obtaining, cooking methods and second time usage of food waste.

无菜不欢，喜吃鱼

广东人酷爱吃青菜，尤其偏爱当地出产的时令青菜。在广东有这样一句调侃："三日不吃青，走路不正经。" 就像四川人和湖南人无辣不欢那样，广东人无青菜不成宴，再怎么吃喝也感觉少了些欢乐，仿佛体内少了些支撑。一位在广东居住多年的外国外交官朋友曾打趣道："如果在家宴请广州人，无论准备多少美味佳肴，都一定要有一盘青菜，否则盛宴便显得不完整了！" 一语道破了老广们 "无蔬菜不成宴"的饮食嗜好。

广东人的餐桌上，常见的蔬菜种类丰富多样，几乎可以说一年四季都能吃到新鲜的时令蔬菜。春天有荿麦菜、春笋、芦笋、青椒和芹菜；夏天有芥蓝、丝瓜、苦瓜、冬瓜、苋菜和空心菜；到了秋天，又有番薯叶、秋葵、豌豆、四季豆、豆角和上海青；而冬季则有菜心、卷心菜、奶白菜、菠菜、芥菜和莴苣。无论是从菜市场采购，还是自己在家栽种，广东人对蔬菜的新鲜度和嫩脆程度都有极高的要求。烹饪时，多采用白灼、氽烫或快速清炒等便捷的方式，以保留蔬菜的原汁原味。

除了蔬菜，广东人还喜欢吃鱼，不论是淡水鱼，还是海鱼，都是广东人餐桌上的"常客"。常见的品种有海鲈鱼、塘饲养鲈鱼、黄花鱼、带鱼、多宝鱼、比目鱼、鲳鱼、鲩鱼、马鲛鱼、石斑鱼、海鳗鱼、叶子鱼、赤鲸鱼和三文鱼等。烹饪的方式多用清蒸、焖、红烧、煎或者做成鱼生（脆肉鲩）。而在享有"世界美食之都"美誉的顺德，人们更是将鱼的各个部位运用得淋漓尽致：一条普通的大头鱼，经过顺德厨师的巧手，可以变幻出数道佳肴，从鱼头、鱼背、鱼腩、鱼骨到鱼皮、鱼尾，毫不浪费，令人叹为观止。

广东人对蔬菜和鱼的热爱，已成为他们饮食文化中不可或缺的一部分。无论是清淡的蔬菜，还是鲜美的鱼肉，都能体现出老广们对自然食材的尊重和对美食的追求。

Can't Go Without Vegetables and Love for Fish

Guangdong people love to eat green leafy vegetables, and most of these are seasonal vegetables produced locally. There is a joke in Guangdong that goes "One should start to have trouble walking straight if one goes without green vegetables for three days". Just like people in Sichuan and Hunan are grumpy without spicy food, Guangdong people would become cranky without green vegetables. No matter how much other foods are available or how much they drink, they feel less happy and less supported. Once a diplomat friend of mine living in Guangdong said that after living in Guangzhou for some time, he has come to the conclusion that if he entertained Guangdong people at home, no matter how delicious the dishes, he should always have a dish of green vegetables. Otherwise it'd be an incomplete meal! "No vegetable, no feast" and so be it.

There are a couple dozen of vegetables amongst

the usual diet of Guangdong people. Fresh seasonal vegetables can be found all year round. In spring, you can eat Chinese romaine lettuce, spring bamboo shoots, asparagus, green pepper and celery; In summer, have Chinese kale (Gai Lam), loofah, balsam pear, winter melon, amaranth and Chinese cabbage; In autumn, you can find sweet potato leaves, okra, pea pods, green beans, Chinese peas and baby Shanghai Bak Choy; In winter, there is Choy Sum, Chinese beet root, Bak Choy, spinach, mustard leaf and lettuce. Either buy from the nearby vegetable market, or simply grow them at home. It is easy to get these ingredients. Guangdong people pay attention to the freshness, tenderness and crispness of green vegetables. The best way to cook them as to preserve the natural tastes of ingredients are steaming, blanching, scalding and quick stir-frying.

On the other hand, Cantonese like to eat lots of

fish, freshwater or from the sea. Common fish varieties on Cantonese dinner tables include sea bass, various kinds of carp (black carp, silver carp, bighead carp), pond raised bass, yellow croaker, hairtail, turbot, flatfish, pomfret, grass carp, mackerel, grouper, sea eel, red croaker, and salmon. Steamed, braised in soy sauce, fried or prepare the fish into raw fish are often used for cooking. In particular, Cantonese chefs are good at making full use of all parts of the fish. In Shunde—"City of Gastronomy", a bighead carp can be put into the hands of a Shunde chef to create countless different dishes. You can eat it from head to tail without wasting anything. The fish head, fish back, fish belly, fish skin, fish tail, any part can become a delicacy, which is simply amazing.

The Guangdong people's affection for vegetables and fish has become an indispensable part of their culinary culture. Whether it's the simplicity of fresh vegetables or the delicate flavors of tender fish, these ingredients reflect the Cantonese's respect for natural produce and their pursuit of delicious good food.

广式烹调中的低碳与节能之道

粤菜饮食文化博大精深，傲古骄今。自古便有"生在苏州，住在杭州，食在广州，死在柳州"之说，足见粤菜在中国美食中的重要地位。粤菜发源于广东，兼收并蓄、博采众长，有"杂而精巧"的特点。据一位入行多年的粤菜名师介绍，单论粤菜饮食文化中的烹饪技法便多达70多种，许多技法都是广东地区独有。在这些技法中，不乏操作简便、省时、健康低碳的烹饪方式，这些恰与当今倡导的"绿色低碳生活方式"不谋而合，且适合日常烹调之用。它们也是本书的关注重点。

例如，广东家家户户都熟悉的白灼做法。白灼是指将食材放入沸水中快速煮熟后捞起，时间极短，能有效保持食材的原有质地，令其鲜、脆、嫩的特点得以保留。这种方法非常适合烹饪绿色蔬菜和河鲜、海鲜类食材，食物出锅后只需略加酱油调味，便能品尝到原汁原味的鲜美。此外，广东人常用的清蒸法，也是粤菜中的经典技艺。清蒸源自中国古代，利用蒸汽热量将食材烹熟，最大限度地保留了食物的汁水和鲜嫩口感。类似的低碳烹饪技法还有清炒、做成鱼生和炖煮，它们皆以省时省力、简单快捷为特点，十分适合现代快节奏的生活方式。

粤菜的大师们常常结合两种或多种烹饪技法，力求将食材的风味发挥到极致。例如，"飞水"与"快炒"的结合便是一种省时高效的做法：先将食材放入沸水中快速余烫，以去除其涩味和农药残留，这一步称为"飞水"，然后再将食材放入锅中进行快速翻炒。飞水的步骤不仅能去除杂味，还能避免食材直接下锅翻炒时因受热不均而出现焦黄或发黑的情况。这样的处理方式不仅能够保持菜品色香味俱全，更是粤菜饮食文化中色、香、味俱全的法宝。

The Green and Low-Carbon Ways in Cantonese Cooking Methods

Cantonese cuisine culture is extensive and profound. It has made food lovers proud since ancient times and in the present. There is a saying that goes"(It is better to be) born in Suzhou, live in Hangzhou, eat in Guangzhou, and die in Liuzhou", which shows the important position of Cantonese cuisine in Chinese cuisine. Cantonese cuisine originates from Guangdong, embodies a "diverse yet refined" characteristic, blending influences from different regions and is a symbol of "miscellaneous and delicacy". According to one famous Cantonese chef with years of experience, there are over 70 kinds of distinct cooking techniques in Cantonese cuisine, many of which are unique to the region. Among these cooking techniques are a variety of methods that are simple, time-efficient, healthy, energy-saving and low-carbon. These techniques are also among the focus of this book, great for daily use and in line with the

"green low-carbon lifestyle" advocated by our society today.

For example, every household in Guangdong is familiar with blanching (bai zhuo). Bai zhuo, involves briefly cooking food ingredients in boiling water, allowing them to retain their natural texture and freshness. The time taken is as little as it can be and it is suitable for cooking green vegetables, freshwater products or seafood. After the food is cooked, add soy sauce to add some extra tastes. Another example of quick cooking is the steaming method commonly used by the Cantonese. This technique of steaming originated from the ancestors of China. It cleverly uses hot steam flow to cook food, while retaining the juiciness of the meat to the greatest extent. In addition, there are also the time-efficient and labor-saving cooking methods such as quick stir-frying, raw fish slicing and stewing, all equally quick and easy to execute, making them ideal for modern, fast-paced lifestyles.

Some Cantonese chefs combine two or more of these cooking techniques to give full play to the flavor of the ingredients. For example, the combination of "feishui" and "quick stir-fry" is time-saving and easy to operate. First, put the ingredients into boiling water and drag them across the water quickly to remove their astringent odor and pesticides residing on the skin, then put them into a pot and add spices for a quick stir-frying. The process of "feishui" can effectively avoid the vegetables turning yellowish grey and looking a little like "bruised" when stir-frying. This hallmark Cantonese cooking technique serves as a perfect tool to give color, aroma and taste to your dish.

勤俭节约与食材的二次利用

在许多外地人的眼中，广东人往往给人一种"低调务实、不讲排场、勤俭节约"的印象。广州人喜欢喝早茶，去茶楼时点上"一盅两件"——一壶茶加两份点心，就能消磨一整个上午。老广们认为，"饮茶"的精髓在于聚会和闲聊，至于吃多吃少，完全凭个人喜好。与亲朋好友外出就餐时，最常听见的一句话便是"食几多，点几多"（吃多少就点多少，吃完了不够再加点）。在饱餐一顿过后，若有剩菜，又会被奉行"食得唔好嘥"（能吃则不要浪费）的广东人统统打包带走，刚巧解决下一餐的需要。

此外，自古便有"要娶就娶广东媳妇"的说法。广东妇女以其精明持家而著称——"知悭识俭"（勤俭节约），仿佛天生就有慧眼巧思，善于利用食材的每一部分，甚至能巧妙地变废为宝。最典型的例子便是广东主妇对各类果皮、瓜皮的二次利用，这在全国范围内都是少见的做法。广东地处亚热带，盛产各类新鲜水果，那些在别处被视为垃圾的果皮、瓜皮，放在广东妇人手中瞬间成了宝贝。举例来说，许多人家将柑橘皮晾晒于阳台或是院子中央，制成陈皮后用来泡水、泡茶，甚至炖汤。不仅如此，许多瓜果皮中富含天然植物化合物，具有降血压、清热解毒的功效。挂在阳台一角晾晒的柑橘皮，从远处看还像是一串串橘红色的装饰，为家中增添了温馨的氛围。

广东人巧妙利用食材的传统以及勤俭节约的美德，也为现代社会的可持续生活方式提供了生动的范例。

Culture of Thriftiness and Second Time Use of Food Waste

In the eyes of many people from other provinces than Guangdong, the Cantonese are seen as "low-key, pragmatic, not fastidious about ostentation, and in fact they are quite thrifty". This is perfectly illustrated in the example of Guangdong people going Yum Cha in the morning. Your average Cantonese person walks into a Cantonese restaurant in the morning, orders a pot of tea, and two small dishes of dim sum, they can usually sit with this much for the whole morning. The classic Cantonese believe that what "Yum Cha" really should focus on is the getting-together and chatting other than the food. Ordering more or less it should depend on the occasion and the people's own preference. When I go out with relatives and friends, the most common line I hear is "only order how much you can eat". After a full meal, if there are leftovers, they will usually be packed home by the Cantonese who follow the principle of "don't waste it if you can eat it", which also can save them from doing some cooking for the next meal.

On the other hand, a saying that "if you want to marry, you should marry a Guangdong girl" has been around for since I can remember. Guangdong women "know how to live thriftily". They also seem to be born with smart eyes and thinking, and know how to use every part of a food to the best of their ability and not let anything go to waste. The most obvious example is the reuse of various kinds of fruit skins by local Cantonese housewives, which is rare even for people living in other parts of China. Guangdong has a subtropical climate and produces abundant fresh fruits. The fruit and melon skins that are disposed of elsewhere or seen as garbage suddenly turned into treasure when they are put into the hands of Guangdong women. In Guangdong, it is common to see local housewives drying mandarin peels on the balcony and in the open space of their backyard, and then use the food scrap to make all kinds of drinks, tea or even soup. The mandarin peels sometimes hung along the window frame in people's homes are seen as strings of orange color decorations that lightens up a home.

This tradition of second time use of food wastes and practicing frugality by the Guangdong people both serve as vivid examples for sustainable living in today's modern world.

无公害种植香料与时蔬

有人曾说："广州是一片插筷子都能发芽的土地。"的确，凭借优越的地理环境和肥沃的土壤，这里能种活大多数植物。由于种植简单，不少广东家庭利用阳台或后院的空地种植蔬菜和香料，依赖大自然的滋养，几乎无须施肥便能收获满满。

在粤菜中，日常使用的葱和蒜就非常适合家庭种植。比如，葱只需将用剩下的根段插入花盆中，露出一点点根部在土壤之外，浇水后不久便会发芽、长出新葱。而大蒜则可以将发芽的蒜瓣剥皮后埋入土中，等到小芽长高后，移到阳光充足的地方，每天浇水，10天左右就能收获新鲜的大蒜了。此外，一些香草如芸香草、迷迭香等，不仅能为日常烹饪提供新鲜调料，还能为家居增添几分绿意和生机。

广州被誉为中国的"食材之乡"，这里适合种植的香料和蔬菜种类繁多，数以百计。香料香草包括芸香草、薄荷、百里香、罗勒、香茅、茴香、花椒、孜然、丁香、草果、桂皮等；而常见的蔬菜有菜心、芥蓝、芥菜、荠菜、番薯叶、青瓜、南瓜、豆角、番茄、苋菜和空心菜等。

自家栽种香草与蔬菜，早已成为广东人生活中的一项优良传统。实际上，自从现代农业普遍使用化肥代替传统有机肥后，许多蔬菜逐渐失去了昔日的爽脆与清甜，反而带上了一丝涩味。通过家庭种植的方式，不仅避免了农药的使用，使蔬菜口感更为纯正，还能减少生产、包装和运输过程中排放的二氧化碳和其他温室气体，为环保事业贡献一份力量。

自家种植不仅是一种返璞归真的生活方式，更是一种绿色环保的实践。广东人通过这项传统，不仅品味到了天然的食材，也传递了对自然与生活的热爱。

广州人家窗台上种的芸香草与罗勒叶
Pots of common rue and basil growing right on the windowsills of homes in Guangzhou

Planting in Your Home Seasonal Vegetables and Herbs

Some people say that "Guangzhou is a land where even chopsticks stuck in the soil can sprout". Indeed, with thanks to its unique geographical location and fertile soil, most things you put in the soil will grow. Because it is easy to grow things, many Guangdong people plant vegetables and herbs on the balcony or their own backyard. Waiting for the wind and rain of Mother Nature, and you can harvest without fertilization.

The shallot and garlic commonly used in Cantonese cooking are perfect for home planting. Insert the root pieces of a shallot into a clay pot with soil, leaving some segments of the root above the soil, exposing them to the sun and with some drops of water, soon they will sprout. You will have new shallots. As with garlic, peel off the pieces that had sprouted and were supposed to be abandoned, and bury them in the soil. After the sprouts grow tall, move them to a sunny place to be watered every day. After more than 10 days, new garlic can be picked. Some herbs, such as common rue and rosemary, are also popular choices. Planting them in your home not only makes it convenient for cooking anytime, but it also helps to add aesthetic vitality to the home environment.

Some people call Guangzhou the "hometown of food materials" in China. Herbs and vegetables that can be planted here are found in the hundreds. Common spice herbs include common rue, mint leaf, thyme, basil leaf, lemongrass, fennel, Chinese prickly ash, cumin, clove, nutmeg and cinnamon. Popular vegetables include Choy Sum, Chinese kale, mustard, sweet potato leaf, cucumber, pumpkin, bean, tomato, amaranth, and Chinese cabbage.

In Guangdong, it has become an excellent tradition to grow vegetables and herbs at home.

In fact, since farmers switched to using chemical fertilizer in large quantities and eliminated the traditional organic fertilizer for growing vegetables, the taste of vegetables has gradually lost its formerly crispness and sweetness and become a little astringent. Planting your own vegetables without using pesticides would not only make them taste better, but also reduce the carbon dioxide and other greenhouse gases emitted in the process of food packaging and transportation, helping to protect the environment.

Home planting not only brings one closer to nature, but is also a sustainable way of living. Through this practice, Guangdong people get to enjoy the freshest and most natural ingredients while showing their love for life and respect for the planet.

绿色出行——集购物、休闲与运动于一身
Getting around the green way—combining errands, fun, and exercise all in one go

第 2 章

Chapter 2

国外饮食中可借鉴的绿色与低碳

Green and Low-Carbon Living in Other Food Cultures

广东人对于来自世界各地的美食一向采取包容开放以及跃跃欲试的态度。正因如此，过去三十年间，广东各地涌现出众多国际美食餐厅，并得以在这片美食沃土扎根下来。"食在广州，美食天堂" 并非随便说说而已，广东的餐饮文化展示的是一种海纳百川的气势。

在本章中，我特别选取了一个亚洲国家和三个西方国家——日本、法国、以色列和德国。这四个国家在低碳与环保方面走在世界的前列，其饮食文化和生活习惯都深受这一理念的影响。我们将探讨他们的饮食文化中有哪些低碳环保的做法值得我们借鉴。在广东，也能找得到提供这四国地道美食的餐厅。

一个有趣的现象是，这四个国家的餐饮中，蔬菜沙拉似乎是必不可少的开胃菜。而且在这些国家，蔬菜沙拉多以生食为主，旨在最大程度上保留蔬菜的原始风味与营养。我想，大概没有人会反对，食用蔬菜并采用简约的烹饪方式，除了能够省时省力，还能有效减少碳排放量，减轻人类对环境造成的负担。

Guangdong people are known for being adventurous when it comes to food and they have an accepting, and sometimes even eager, attitude towards trying foods from abroad. Because of this, over the past 30 years, a large number of international food restaurants have sprung up all over Guangdong, and have been able to thrive. The Chinese saying "Eat in Guangzhou, a Food Paradise" was not created by accident. Just as "all rivers run into the sea (海纳百川)", so does Guangzhou, food paradise of China, incorporate hundreds of different culinary traditions into its rich and deep food culture.

In this chapter, I have chosen to spotlight one other Asian nation—Japan, and three Western nations—France, Israel, and Germany. All four are at the forefront of advocating for low-carbon living, and their food cultures and living habits have been affected by their efforts. In this chapter, I will share some of the green and low-carbon aspects of their cuisine and cultures. In Guangdong, you can also find restaurants dedicated to serving authentic cuisine from each of these four countries.

One interesting characteristic that seems to be shared amongst these four nations is that salad is an indispensable appetizer. Vegetables found in these salads are mostly eaten raw, in order to preserve the natural tastes and nutrients of the vegetables. Eating vegetables and adopting simpler cooking methods not only saves time and reduces the burden of food preparation, but also produces less carbon emissions, which in turn helps ease the pressure on the natural environment.

日本

在日本的烹饪哲学中，讲究的是最大限度地减少人为干预，强调尽可能保持食材的自然状态。这种理念极致地表现在"最好生吃，其次烤了吃，只有食材不适合生吃或烤了吃时，才选择煮熟了吃"。这也许可以解释为何刺身能常年不衰稳居日本料理菜单的"王位"。少用复杂的烹饪技艺和调味，不仅减少了烹饪过程中的能源消耗和资源消耗，有助于降低温室气体的排放，还能节省时间，这种低碳、绿色的烹饪方式值得借鉴。

日本传统饮食以鱼、米饭和蔬菜为主，且日本人有"不时不食"的说法。日料厨师们根据时节安排蔬菜，尤其偏爱各种食用菌。常见的日本菇菌类包括香菇、金针菇、灰树菇、杏鲍菇、滑菇、蟹味菇、茶树菇、猴头菇以及野生松茸等。这些食用菌的做法多样，有烧烤、天妇罗、炙烤，或者切片与蒸蛋一同制作成"茶碗蒸"。

食用菌被视为自然界的"垃圾分解员"，具有化腐朽为神奇的本领。在自然界中，它们通过分解枯木起着重要的生态作用，如果没有食用菌，地球将被腐木覆盖，难以维持生命。多吃食用菌不仅有助于清理地球上的自然垃圾，还能保护环境。自从日本学者北本（Kitamoto）等人培育出世界上第一个白色金针菇品种以来，日本在食用菌的科研和设施栽培技术方面一直处于世界领先地位。

值得一提的是，现今中国已快速发展为全球食用菌生产大国。无论是本土发现的种类，还是从国外引进的品种，中国的食用菌总产量稳居世界第一，种植技术水平日趋成熟。除了培育出诸如白灵菇、猴头菇等自主品种外，中国还向日本出口真姬菇和野生松茸等食用菌类。

Japan

In traditional Japanese cooking philosophy, there is a strong emphasis on minimizing the amount of cooking in order to preserve the freshness and natural flavors of the ingredients. There is even a hierarchy in Japanese culinary cuisine that classifies food from the most fine dining and exquisite to lesser grades based on how it is prepared. "The best is raw food, followed by roasted food. If the food cannot even meet the standard to be roasted, then cook it" says a famous Japanese culinary master. This philosophy may help explain why sashimi occupies the throne of the Japanese menu all year round. De-emphasizing complex cooking techniques and seasoning reduces the energy required for the cooking process, which in turn contributes to fewer greenhouse gas emissions, and also saves time.

The traditional Japanese diet includes fish, rice and vegetables as the main ingredients, and there is a popular saying that goes "Not in season, don't eat it (不时不食)". On an average day, chefs use only seasonal vegetables, and there is a high preference for mushrooms. The varieties of mushrooms and fungi typically found on the Japanese table include shiitake, needle mushroom, maitake, apricot abalone mushroom, nameko, beech mushroom, tea tree mushroom, hericium erinaceus and wild matsutake. The cooking methods include braising, tempura (frying), baking or slicing, as well as use in Japanese savory steamed custard.

The growing of edible fungi and mushrooms are part of a process used by Mother Nature to facilitate the decomposition of matter. It has the power to "transform decay into magic". For example, edible fungi participate in the decomposition process of rotten wood, without it, the earth would be covered by rotten wood and could not survive for long. Growing more edible fungi can help clean up the environment. Since Kitamoto, a Japanese scholar, cultivated the first white needle mushroom variety in the world, Japan has been at the forefront of the cultivation technology of edible fungi.

It is worth mentioning that China has also rapidly developed into a major producer of edible fungi in recent years, including those found locally. The total production of edible fungi in China now ranks first in the world, and its planting technology is maturing. In addition to cultivating its own varieties of edible fungi, such as Bailing Mushroom and Hericium erinaceus, China is now even exporting agaricus and wild matsutake to their original birth place, Japan.

法国

在法国，有机食品向来备受推崇。有机食材的生产和加工过程受到严格监管，以确保种植土地未使用化学肥料，并遵循环境保护原则，如休耕轮作、动物福利保护等。此外，有机食品的原料还不能含有香味添加剂、色素或合成化学香料。法国街头随处可见专门销售有机食品的商店，甚至一般超市中也设有标有"Bio"标识的有机食品专卖区。许多法国人认为，尽管购买有机食品可能增加开支，但其优势显而易见：不仅能享受到更高品质、更新鲜的健康食材，还能保护土壤与生态平衡，减少农药和食品添加剂对环境及人体健康的侵害。

在绿色生活方式方面，法国政府对公共交通基础设施进行了大量投资，重点发展电动车和公共汽车，并为购买电动或混合动力汽车的消费者提供税收优惠。政府还制定了到2040年停止销售汽油和柴油汽车的目标，以减少空气污染和碳排放量。这些举措对环境保护具有积极的推动作用。

我曾有机会去法国拍摄节目，在一个周末的清晨，时常能见到悠闲骑着自行车去集市购物的市民。他们将购买的面包或者鲜花放在车筐里，穿梭于城市的大街小巷之间。这种生活方式将购物、运动和绿色出行完美结合，可谓一举三得，体现了法国人对环保与健康生活的追求。

France

In France, organic food has always been popular. The production and processing of organic ingredients are strictly monitored by relevant authorities, ensuring that no chemical fertilizers are used on the land, the environment is respected by allowing fields to lie fallow, animals are protected through the ban of animal testing, etc. Additionally, to qualify as organic, food ingredients must not contain flavor enhancers, colors, or synthetic chemicals. You can find stores selling 100% organic food all over France. Even regular supermarkets have dedicated sections for organic products marked with a "Bio" label. Many French people believe that although buying organic food increases daily expenses, the benefits are clear: consumers can enjoy higher quality, fresher-tasting, healthier food, while contributing to the protection of soil, greater ecological balance, and a reduction of the harm caused by pesticides and food additives to both the environment and the body.

France is also at the forefront of sustainable transportation. Its urban planning models are very pedestrian and bicycle friendly. The local government has heavily invested in the public transportation infrastructure, focusing on electric vehicles and buses. Additionally, tax incentives are provided for purchasing electric or hybrid cars, with a goal to stop selling gasoline and diesel cars by 2040, helping to reduce air pollution.

During my time in France filming a TV program, I recall how on weekend mornings I saw people leisurely cycling to the local markets. Toting baskets filled with freshly baked bread or flowers, they pedaled through the streets and alleys, seamlessly blending shopping, exercise, and eco-friendly transportation. It was a role-model illustration of how to achieve multiple benefits in one elegant move.

以色列

在以色列，当地人对水资源的节约与高效利用可以说走在世界前列。水是生命之源、万物之基，没有水，就无法耕种，粮食也无从谈起，更别说维持生命了。以色列位于中东沙漠边缘，处于欧、亚、非三大洲交会处，大部分国土为沙漠，自然资源极为匮乏，人均水资源仅为270立方米，相当于中国人均水资源的八分之一。然而，短短几十年间，以色列迅速发展成为全球科技农业强国，沙漠中出现片片绿洲，种植出新鲜的水果蔬菜，其国内粮食生产基本实现自给自足。这一成就得益于两个关键因素：一是政府营造的全民节水氛围，二是以色列科学家通过创新发明，最大化地利用水资源进行农作物灌溉。

截至2024年，以色列家庭用水量位居全球最节约国家行列。政府通过电视、报纸等媒体长期宣传"水贵如油"的理念，提醒民众珍惜水源。许多家庭在自家屋顶安装了雨水收集系统，将雨水储存在水桶或沟槽中，用于花园浇灌。另外，以色列人使用过的水不会轻易倒掉，而是循环再用，如洗车、浇花或存储等待回收。

以色列科学家还开发了先进的滴灌技术，以减少水资源浪费。该技术根据作物的需水量进行精准灌溉，比如只湿润作物根部附近的土壤，每次灌溉量都经过精确控制，从而实现持续灌溉，避免浪费。如今，在以色列随时都能品尝到新鲜的水果和蔬菜。最有名的水果包括鳄梨、雅法橙和出口欧美的柚子等。

在饮食习惯方面，以色列人热爱吃沙拉。餐桌上经典且常见的是以色列沙拉，它的食材在当地容易获取、操作便捷：将青瓜、青红椒、洋葱、香菜和番茄等切成细小颗粒，拌入橄榄油、黑胡椒和盐调味即可。一位生活在广州的以色列犹太朋友告诉我，在以色列，几乎每家每天都会吃这种以色列沙拉。来到广州后，由于食材易得、制作简单，她每天也会用广州当地的蔬菜为家人制作以色列沙拉，既健康又低碳。

Israel

In Israel, the conservation and efficient use of water resources is said to be at the forefront in the world. Water is the source of life and the foundation of all things. If there is no water, there will be no crops, no food, and no life. Israel is located at the edge of the desert in the Middle East and at the junction of Europe, Asia and Africa. Most of its land area is desert and natural resources are scarce, the per capita water resources are 270 cubic meters, which is only 1/8 of China's per capita water resources. However, it has rapidly developed into one of the global scientific and technological agricultural powers in just a few decades. Oases have appeared in the desert, fresh vegetables and fruits have grown, and Israel has become self-sufficient in its food production. This could be attributed to two key factors: first, the local government created a culture for saving water amongst its citizens, and second, through innovation and technological invention, water resources can be maximized for agricultural irrigation.

To this date of 2024, the domestic water consumption in Israel is one of the most economical in the world—the local government constantly publicizes that water is as expensive as oil through television and print media, reminding people to treat water resources well and develop good water-saving habits. For example, some households install rainwater collection systems on their roofs, collecting rainwater in buckets for watering gardens or cleaning. In addition, the Israelis will not dump the water soon after using it, but recycle it, using it to wash cars, water flowers or storing it for waste water recycling.

In order to reduce water loss, Israeli scientists have developed advanced drip irrigation technology. This technology makes sure that irrigation is timely and appropriately conducted according to the water demand of the crops, for example, only the soil near the roots of crops is wetted and the amount of water for each irrigation is precisely controlled, thus achieving continuous irrigation and avoiding waste. Thanks to the world's leading agricultural drip irrigation technology, Israelis can eat fresh fruits and vegetables anytime and anywhere. The most popular local fruits also include avocados, Jaffa oranges and grapefruit which are popular in Europe and America as well.

In terms of dietary habits, the Israelis are known for their love of salads. The most common vegetable dish on the table of local Israeli people is the Israeli salad, which is eaten raw and made of easy-to-obtain ingredients. Cut the cucumber, green and red pepper, onion, coriander and tomato into small pieces, stir them together, and then add olive oil, black pepper and salt to taste. An Israeli friend living in Guangzhou once told me that almost every family in Israel has Israeli salad every day. After coming to Guangzhou, because its ingredients are convenient to obtain in Guangzhou, she makes it every day for her families.

德国

德国人非常注重节省时间，大多数人不希望在烹饪或用餐上花费太多精力。对于饮食，德国人或许不如其他国家那样走心，但他们却十分重视营养搭配和膳食均衡。一般来说，早餐是德国人一天中最为丰富的一餐，通常包括主食、肉类、水果、蔬菜、牛奶、果汁和咖啡等。至于午餐，许多德国人会选择在公司食堂或者快餐店快速解决：一份土豆泥搭配沙拉，再配几块肉类或三明治加饮料，简单快捷。如果在家吃午餐，主妇们则常常会煮一些熟肉，搭配水煮蔬菜，饭后再来一杯咖啡。而晚餐方面，不少德国人习惯吃"冷晚餐"，即直接食用无须烹饪加热的食物，直接从冰箱中取出即可。

几位我认识的德国厨师以及德国驻穗外派人员曾与我分享过典型德国家庭的"冷晚餐"构成，通常是黑面包搭配冷肉类，如熏牛胸肉、猪肝肠、萨拉米香肠、里昂那香肠或火腿片中的任意两三种；再切一些生的青椒或黄椒条，配上腌酸黄瓜、酸蘸头和腌制的马鲛鱼，就是一顿简单的晚餐了。这样的"冷晚餐"不仅节省了烹饪时间，还减少了能源消耗。一位定居广东的德国厨师告诉我，德国丈夫们在外工作一天后通常希望回家见到心情愉悦的妻子，而便利的"冷晚餐"除了环保，也减少了妻子们准备食物的负担，有助于其保持心情舒畅，促进夫妻间关系和睦。

为了推动环保，德国政府一方面通过调整大学生饮食结构来支持低碳环保，另一方面从2021年7月起禁止销售一次性塑料吸管、食品容器、聚苯乙烯杯子和餐盒等塑料制品，展现出其告别"一次性文化"的决心。

Germany

The Germans attach great importance to efficiency and saving time. It is said most people do not want to spend too much time on cooking or eating. That said, they also pay attention to having a nutritious and balanced diet. Generally speaking, breakfast is the most abundant meal of the day, usually including staple food, meat, fruit, vegetables, milk, juice, coffee, etc. For lunch, most Germans will choose to eat quickly in the company canteen or a nearby cafe. Usually, mashed potatoes and salad are combined with several pieces of meat to form a platter, or sandwiches and a drink are added. When at home, the wife will usually cook some meat and serve them with boiled vegetables, and a cup of coffee is drunken after dinner. It is interesting to know that some Germans have the habit of eating a "cold dinner". You may ask, "Is this dinner without a trace of hot steam?" Correct. It means eating food that does not need to be cooked or heated, often directly taken out of the refrigerator.

I know several German chefs as well as German diplomat friends who share with me the main ingredients of the typical German "cold dinners". It usually includes black bread, served with any two or three kinds of cold meat like beef pastrami, pig liver sausage, salami, lyoner sausages and ham slices. And add some raw green and yellow pepper strips, pickled cucumbers, pickled leeks and marinated mackerel you have a dinner ready. The habit of eating "cold dinners" allows Germans to save more time to do other things and reduce the energy needed for cooking. A German chef who now lives in Guangdong once told me half jokingly that part of the reason why "cold dinners" are popular is that German husbands want to go home to a happy wife who would be happier if she did less cooking and housework. "Happy Wife, Happy Life." he adds. And "cold dinners" do just that!

In addition to supporting low-carbon environmental protection by changing the diet structure of university students, Germany also has banned the sale of disposable plastic straws, cotton swabs, food containers, polystyrene cups, boxes and other plastic products from July 2021. The Germans are determined to say goodbye to the "disposable culture".

我小时候与父亲在我家后院种植蔬菜
When I was little, I'd help my dad grow vegetables in our backyard—it's one of my fondest memories

第 3 章
Chapter 3

低碳取材
Sourcing Ingredients with Smaller
Carbon Footprints

我们知道"碳排放量"一词指的是在生产、运输、使用及回收某件物品时所产生的温室气体排放量。在本章中，我们将重点探讨粤菜中那些在生产和运输过程中碳排放量较低的菜肴和食材。总体来说，蔬菜种植的碳排放量远低于白肉类（如鸡肉、鸭肉）养殖的，而白肉类养殖的碳排放量又低于牛羊等红肉养殖的。有研究表明，蔬菜种植的碳排放量比牛羊肉养殖所产生的碳排放量低了90%。

如果你有幸生活在广东，那么在考虑吃什么的时候你无疑拥有天然的优势，因为这里一年四季都能产出各类蔬菜和水果（关于广东四季蔬菜的详细种类可参考本书第4页），这些蔬菜味道清甜，各具特色，丰富多样。如果喜欢，每天都可以品尝不同种类的蔬菜，不仅满足味蕾，还能为环境保护出一份力。有一些移居海外的广东移民，因为怀念家乡的绿叶蔬菜，还会在当地自己种植。这些蔬菜在当地还因其粤语发音而得以命名，例如在美国、加拿大、澳大利亚、新西兰和秘鲁的超市中，常见出售的菜心叫"Choy Sum"，而芥蓝被称作"Gai Lam"，均是从粤语音译而来。

当然，减少碳排放量并不意味着完全拒绝肉类，毕竟肉类富含铁、蛋白质等对人体有益的营养成分，适量食用肉类是健康饮食的一部分。我们建议日常多吃蔬菜、少吃肉类，特别是在选择肉类时，尽量多选择白肉，减少红肉的摄入。

另外，我也推荐大家在有条件的情况下，在家中种植蔬菜和香草。在广东，许多蔬菜和香草都容易种植成功。本章中我也会介绍一些蔬菜和香草的种植方法。现在市面上许多非有机蔬菜使用化肥种植，使得蔬菜口感不再鲜脆，且带有一丝涩味。而自己种植蔬菜不仅可以避免农药的使用，让食物更加天然美味，还能减少生产、包装和运输过程中产生的二氧化碳等温室气体的排放。另外，种植蔬菜还可以美化家居环境，增加家中空气的含氧量。

We are familiar with the term "carbon emissions", which refers to the average greenhouse gas emissions generated during the production, transportation, use and recycling of any given product. This chapter presents to you classic Cantonese dishes and ingredients that are low in carbon emissions during their production and transportation phases. Generally speaking, the average carbon emissions of vegetable planting are much lower than that of white meat breeding. And the average carbon emission of white meat breeding such as chicken is lower than that of red meat breeding such as lamb and beef. Studies show that the carbon emissions of most vegetables are 90% lower than that of red meat such as beef or lamb.

If you have the opportunity to live in Guangdong, you are very lucky, because there are many kinds of vegetables and fruits that are available all year round. Their flavor is generally sweet, delicious and unique. If you want, you could choose a different variety to eat, every day, 7 days a week, which is not only good for your diet, but also a good choice for the environment. It's really killing two birds with one stone. Some Guangdong immigrants living overseas miss Cantonese green leafy vegetables so much that they plant these vegetables in their own backyard. These vegetables have become so popular overseas that in countries like the U.S., Canada, Australia, New Zealand and Peru, people even refer to them by their Cantonese names; "Choy Sum" and "Gai Lam" are commonly sold in supermarkets in these countries.

Of course, supporting carbon reduction does not mean we should completely rule out eating meat. After all, meat is rich in iron, protein, and many other beneficial nutrients to the human body. As with anything, moderation is key. We recommend incorporating more vegetables and less meat into your daily meals, and when choosing meat, opting for white meat rather than red meat whenever possible.

Generally speaking, I encourage readers to grow their own vegetables and herbs whenever possible. In this chapter, I will also introduce planting methods for these vegetables and herbs. In Guangdong, many varieties can be easily cultivated. The (non-organic) vegetables that can be bought on the market now use chemical fertilizers, and the taste of the vegetables is less crisp and sweet and more astringent. And more often than not, pesticides are used in the growing process. Besides having a better taste, growing your own vegetables at home can also reduce greenhouse gases such as carbon dioxide or carbon emissions and eliminates the need for food packaging and transportation altogether. Furthermore, some can also beautify your home environment and increase oxygen levels.

广东人都爱炒菜心

在所有绿叶蔬菜中，菜心算得上是广东人的最爱。资料显示，广东省内种植的菜心几乎从来不往外省输出，因为仅在广东本地就能够被完全消耗殆尽。由于广东人对菜心的偏爱，有不少商家甚至远赴广西、海南乃至宁夏等地，借助肥沃的土壤和较低的地价种植菜心，再运回广东供给本地人食用。而其中最负盛名的增城迟菜心，更是享有"菜心之王"的美誉。广东人每年要吃掉全国约50%的菜心；在海外的唐人街超市里，菜心也是最常见的绿叶蔬菜之一（因为海外华人中，广东籍贯的移民占据了大多数）。美国、加拿大和澳大利亚等国家甚至为菜心取了个英文名，直接采用粤语的发音——"Choy Sum"。

菜心不仅味道鲜美，还长得优雅美观——它身段柔软，翠绿的茎秆顶端开着一簇金黄色的小花，随风摇曳，仿若百蔬仙子。菜心喜好温和的气候，尤其适合在南方地区栽种。不少乡镇村民会在自家庭院的空地上种植菜心，远远望去，碧绿的菜田点缀着星星点点的黄花，惹人喜爱。

要想吃到一道上好的清炒菜心，选材至关重要。要选择那些梗部呈棱角的菜心，最为清甜爽口，菜梗呈圆形又规则的反而没那么可口。将菜心买回来冲洗干净，用手摘掉菜心靠近茎部较硬的部分和不能轻易折断的老叶，留下只有十几厘米的幼茎和花朵部分即可。

在广东，菜心最常见的做法就是清炒，几滴油、一点盐，便可成就一碟美味。将清洗干净的菜心倒入热油锅中，喜欢的话，可以来点猪油渣子做配料，"用猪油的荤香激活菜心之素"。当油开始微微冒烟时，加入少许盐和糖，再迅速倒入菜心，急翻快炒。菜心下锅后会迅速收缩，颜色变成深绿，所以要把握好炒制的时间：喜欢爽脆口感的，翻炒1分钟即可；若偏爱软嫩的口感，则可以稍加一点冷水，盖上锅盖焖个1分钟左右，再起锅上盘。我曾经听人说清炒菜心的时候要滴几滴绍兴酒，或者加入姜、蒜，但其实这些调料往往会掩盖菜心本身的清甜，得不偿失。

小时候，清炒菜心是我们家餐桌上最常见的青菜。我记忆中最美味的菜心，是那种只放一点盐，其他任何调料都没有的。端上桌时，菜心热气腾腾、翠绿欲滴，夹起一条放进嘴里，鲜甜脆爽，不一会儿工夫一整碟已经下肚。

广式生炒菜心
Cantonese-style stir-fried Choy Sum

Guangdong People Love Stir-Fried Choy Sum

Among all the green leafy vegetables, Choy Sum holds a special place in the hearts of Guangdong people. Interestingly, statistics reveal that Choy Sum cultivated and produced in Guangdong province is rarely exported elsewhere, as its abundance caters perfectly to the local consumption within the region. Due to the Cantonese's strong affinity for this delightful vegetable, numerous retailers even undertake long journeys to Guangxi, Hainan, Ningxia, and other regions to either cultivate or purchase Choy Sum in substantial quantities, which they then transport back to Guangdong for sale to the local people. The most renowned variety of Choy Sum hails from Zengcheng city and is fondly known as Chi Choy Sum. The Cantonese preference for Choy Sum has risen to remarkable heights, with Guangdong people reportedly consuming approximately 50% of the nation's total Choy Sum yield each year. Interestingly, in Western nations, Choy Sum

remains one of the most commonly found green leafy vegetables in Chinatown supermarkets. It has gained such popularity that the governments of the United States, Canada, Australia, and other countries have adopted its Cantonese name and pronunciation—Choy Sum.

Choy Sum is not only delicious but also exudes an air of elegance and style. Its soft, emerald stalks adorned with clusters of golden flowers at the top sway gracefully in the wind, resembling vegetable fairies. This green leafy delicacy thrives in mild climates, making it particularly well-suited for cultivation in Southern China. It is a common sight to witness many villagers planting Choy Sum in the open spaces of their backyards, creating picturesque patches of green and yellow flowers that catch the eye from a distance.

If you want to create a truly delicious Choy

Sum dish, selecting the right ingredients is crucial. Choose Choy Sum with angular stalks at the end of the vegetable stem: these are the sweetest and most tender. Those with round and regular vegetable stalks may not be as pleasing to the palate. Once you have the perfect Choy Sum, use your hands to delicately remove the parts near the stem, as well as any branches and leaves that are not easily broken, leaving only about ten centimeters of tender stems with the delightful yellow flowers on top.

In Guangdong, the most common way to prepare Choy Sum is by stir-frying. With just a few drops of oil and a dash of salt, you can transform this vegetable into a delicious dish. Once you've carefully selected and cleaned the Choy Sum, add some oil to the frying pan. For a more flavorful option, you can use lard dregs as an ingredient, as the saying goes, "use the meat of lard to activate the juicy taste of the vegetable." When the oil starts emitting a light smoke, add a pinch of salt and sugar, and then toss in the vegetable, quickly stir-frying it. Choy Sum tends to shrink rapidly and turn dark green in the heat, so take care to finish the process promptly. If you prefer a crispy texture, fry it for less than a minute; for a softer texture, you can add a little cold water, cover the pot, and let it cook for over a minute before turning off the heat and serving. I once heard that some people add a few drops of Shaoxing wine or ginger and garlic when stir-frying Choy Sum, but in reality, these additions might mask the natural taste of the vegetable. Therefore, I would not recommend using them.

When I was young, Choy Sum was the most commonly eaten vegetable in our family. The best way to prepare it was with a simple stir-fry, adding just a little salt. As it was served, the dish arrived at the table steaming hot and vibrant green. With the first mouthful, its delightful taste would captivate us, and before we knew it, the entire dish would vanish in no time.

省心省事、皇帝也夸的番薯叶

　　我从小长大的家门前，有一小片空地，稀稀疏疏地长着几棵树和些许杂草。有一年夏天，小阿姨将几株炒菜用剩下的番薯叶茎段随手扔进了这块土地，任它自生自灭。没想到，这些番薯叶茎段遇土生根，很快形成根系，短短半个月，那原本光秃秃的角落竟然生长出了一片碧绿繁茂的番薯叶。从那之后，我们家开启了在这片空地上种植番薯叶的传统。

　　番薯叶生命力顽强，是广东所有绿叶蔬菜中最易存活的品种之一。它的独特之处在于，在南方种植，虽然能年复一年地绿意盎然，却不结番薯。它仿佛是植物界的劳模，默默奉献，却不图回报，这一点十分难得。番薯叶钟情于广东的温暖气候，抗旱耐热，茎叶柔韧、不易折断，适合家庭种植。可选择在春、夏、秋季，将健壮的番薯叶茎段插入湿润的土壤，10天左右便能生成根系，淋一次肥水后，可任由它自行生长，便可源源不断地收获嫩叶。

　　除了易于种植，番薯叶的营养价值也在各类绿叶蔬菜中名列前茅。亚洲蔬菜研究发展中心已将番薯叶列为高营养蔬菜之一，其富含蛋白质、胡萝卜素、各类维生素以及钙、铁、磷等矿物质元素。与菠菜、生菜、芥蓝、胡萝卜、青瓜等10多种日常蔬菜比起来，番薯叶在14种营养成分含量中都排行第一。难怪香港人称番薯叶为"蔬菜皇后"，美国、日本等地更将其视为"长寿菜"，番薯叶甚至被制成航天食品，供宇航员食用。

　　番薯叶最好吃的做法当然是用蒜清炒。将新鲜采摘的番薯叶清洗干净，蒜头拍碎，剁成蒜蓉，然后开大火热锅，放入足量的油和蒜蓉，再将番薯叶迅速翻炒。粤语中有句俗语说："油多不坏菜，礼多人不怪。"这里的多油能帮助中和番薯叶略带粗糙的口感，增添香气。绿油油的番薯叶吃进嘴中，带着田野的清香。如今，番薯叶已成为南方餐桌上的佳肴，无论是寻常百姓家，还是酒楼餐厅，都少不了它的身影。相传乾隆皇帝南巡时，偶然品尝到了一道番薯叶菜，吃罢龙颜大悦，随即询问菜名。厨师见是皇上问便不敢说是"番薯藤"，灵机一动，称其为"龙须菜"。乾隆闻言欣然点头，从此，番薯叶便多了一个雅号——"龙须菜"。

蒜蓉炒番薯叶
Garlic stir-fried sweet potatc leaves

The Emperor's Love for Sweet Potato Leaves

In the home I grew up, there was a small piece of vacant land in front of the house, with several small trees and some weeds growing on it. One summer, the Ayi(阿姨)after making a stir-fry vegetable dish, threw some leftover raw sweet potato leaves and stems onto the ground. Who would have guessed that these sweet potato leaves and stems would quickly take root in the soil, and half a month later, the originally bare "corner" of land was covered with new sweet potato leaves. Since then, our family has kept the tradition of growing sweet potato leaves on this plot of land.

Sweet potato leaves are easy to grow, probably the easiest among the major Cantonese leafy greens. They've got a soft spot for the southern China soil, especially here in Guangdong and the nearby provinces. Their leaves stay green pretty much all year round, even though they don't actually produce any sweet potatoes. It's

like they're the silent hard workers of the plant world, giving it all but not asking for anything in return. That's a real rarity. sweet potato leaves like the warm weather of Guangdong, they're resistant to high temps and drought, and their stems and leaves don't break easily, making them perfect for the family garden. You can plant sweet potato leaves in spring, summer, or autumn. Go for sturdy stem cuttings to start growing them. Keep the soil nice and damp as they grow. Around 10 days after planting, the roots should be set, and you can let them do their thing. Give them a spray of fertilizer and water, and just keep picking those delicious leaves.

Besides being easy to grow, sweet potato leaves have one of the highest nutritional values among all the green leafy vegetables. sweet potato leaves are rich in protein, carotene, vitamins, calcium, iron, phosphorus and other mineral elements. Compared with more than 10 kinds of daily vegetables, such as spinach, lettuce, Chinese kale, carrots, and cucumbers, sweet potato leaves rank number one with 14 different kinds of nutrients. No wonder Hong Kong calls sweet potato leaves the "vegetable queen". The United States, Japan and other places also regard sweet potato leaves as robust vegetables; they have even been used to make food for astronauts.

My favorite way to cook sweet potato leaves is to stir fry them with garlic. Pick the vegetables and clean them thoroughly. Mince the garlic. Fire up the stove and heat your trusty frying pan. Pour in a bit of cooking oil and your minced garlic, then put the sweet potato leaves in, stir and turn quickly. When considering how much oil to use, there is a saying which goes "more oil will not spoil the vegetables, and anyone can live with more courtesy(gifts)". Oil helps balance out the sweet potato leaves' slightly rugged taste. I'm not exactly sure when sweet potato leaves became a local delicacy in Southern China, showing up on family tables, small eateries, and even high-end restaurants, but legend has it that even Emperor Qianlong couldn't resist their charm. The story goes that he tasted the vegetable while on a trip in Jiangnan and asked the local people for the name of the vegetable. When the chef saw that it was the emperor who had inquired he was afraid to say it was just ordinary sweet potato leaves, so instead he gave them a nickname: longxucai, or "dragon's beard leaf". In the tale, Emperor Qianlong was pleased, and the yam leaf dish is also called the dragon's beard leaf dish to this day.

中秋佳节不容错过的粤菜佳肴

在广东，素来都有在中秋佳节吃紫苏叶炒田螺的习惯。清末民初时期的一首竹枝词写道："中秋佳节近如何？饼饵家家馈送多。拜罢嫦娥斟月下，芋头啖遍又香螺。"这几句诗生动描绘了广东人在中秋佳节时的饮食习俗，而其中最具代表性的，便是这道紫苏叶炒田螺。

炒田螺的历史在广东由来已久，许多大排档与粤菜餐厅都能见到它的身影。最早的时候，人们炒田螺并不用紫苏叶，而是用蒜头、辣椒、薄荷叶和生姜等调料去炒，但是却无法将田螺的泥腥味去除。后来，偶然有人将不起眼的紫苏叶和田螺一起炒，发现紫苏叶不仅能去除田螺的泥腥味，还能为菜品增添独特的清香。从此，紫苏叶与田螺成了天作之合。用紫苏叶炒田螺时，加入适量的盐、料酒和辣椒，成品便有了那种鲜中带甜、甜中带香，并透着丝丝辣意的绝妙风味，让人越吃越过瘾。

紫苏叶是中国南方常见的一种芳香类蔬菜，田野或路边都有生长，叶子呈锯棱状，正面绿色，背面紫色，带有柔软的茸毛。它喜欢温暖湿润的环境，适合栽种在阳台、窗台，甚至客厅等采光充足的地方，播种后不久便能发芽。要想自己种植紫苏叶，可直接将紫苏种子撒在土上，然后用土盖住，浇透水，并在出苗前保持土壤湿润。紫苏从播种到出苗一般需要10～20天，当温度保持在20～25℃时生长比较快。南方不少人将紫苏叶种植在庭院或者客厅中，散发出淡淡的清香。餐厅和大排档甚至会在门口种上几盆紫苏，枝叶层层叠叠，看着生机勃勃，甚是热闹。

小时候，我很喜欢放学后和几位同学结伴去豪贤路上的阿旺大排档觅食。那里的紫苏叶炒田螺是我们最常点的菜品之一，店里经常供不应求。阿旺的老板娘圆乎乎的，每当有客人进店，她总是先眯起那双笑意盈盈的丹凤眼，紧跟着一句："自己揾位坐啦！"（自己找位置坐吧）。厨房里只有她的儿子一个人在忙碌炒菜，而她则兼顾着收银员和女侍应的角色，手脚麻利。各式小吃之中，这紫苏叶炒田螺是他们家最快卖完的菜品，因为田螺拿回来以后需要放入清水中两天，养着吐沙，不能随时补货现做。

有一次，我亲眼看到老板娘将刚买回来的田螺倒入一个大水盆，滴上几滴豆油，注入清水，然后拿着钳子逐个剪掉田螺的尖尾端部分，发出"咔哧咔哧"的声音。刚买回来的田螺有很多细小的沙子，必须用一盆清水养两天让沙子尽量吐清。剪掉了田螺的尖尾端则有利于烹饪时候让调料渗透其中，食用时更加方便嗦出螺肉。

在阿旺吃到的紫苏叶炒田螺，调料讲究，通常会用生姜、蒜头、豆豉、小红辣椒和料酒一同炒制，去腥增香。炒好的田螺肉质丰腴，味道鲜美，清淡爽口，带着紫苏叶的独特芳香。

最好吃的田螺可以到广州郊外山里的小溪或者水田里找。这里的田螺受环境污染小，没有什么沙子且肉质更加爽脆，是大城市里极难享用到的原生态美味。而紫苏叶除了能为田螺增添香味、去腥以外，还有散寒和理气的功效，这道菜也因此被誉为绿色健康的美味佳肴。

今年中秋佳节，月圆之夜，你是否会在厨房炒上一碟紫苏叶田螺，与家人一起，在月光下共享那团圆的温馨时光？

紫苏叶炒田螺
Stir-fried river snails with perilla leaves

清洗田螺
Giving the snails a good rinse before they hit the pan

What the Mid-Autumn Festival Can't Do Without

In Guangdong there is a long-standing tradition of eating stir-fry snails with perilla leaves during the Mid-Autumn Festival. In the late Qing Dynasty to early Nationalist Era, there was a "zhuzhi" poem (poem transcribed from a folktale) that said, during the Mid-Autumn Festival, families send mooncakes to one other, worshipping Chang'e and having banquets under the moon, eating taro root and snails.

In Guangdong, stir fried snails are a very common delicacy that can be found in food stalls and Cantonese restaurants. When I was a child, the most common dish my family cooked to celebrate the Mid-Autumn Festival was stir fried snails with perilla leaves. In my memory, it is the flavor of old Guangzhou. In the early days, Guangdong people did not know to stir fry snails with perilla leaves, but instead used vegetables such as garlic, chili, mint leaves, or ginger, to try and mitigate the muddy and fishy taste of snails. Later, someone happened to fry the inconspicuous perilla leaves and snails together and found that these leaves could remove the fishy flavor. Additionally, adding some salt, cooking wine, and chili peppers would create a unique flavor combination that was refreshingly sweet and sweetly fragrant, with a hint of spiciness. The more people ate it, the more addictive it became. Since then, perilla leaves have become the perfect pairing for snails.

First of all, let's talk about the purple perilla leaf, which is a common aromatic vegetable found in southern China. It grows in fields or by the roadside, with serrated leaves that are green on the front, purple on the back, and has soft hairs. Perilla leaves flourish in a warm and humid environment, and can be grown on windowsills, balconies, and even well-lit indoor spaces. Scatter the perilla seeds on the soil, cover them with soil, water thoroughly and keep the soil nice and damp as they grow. Around 10-20 days after planting, the perilla should sprout. Maintaining a warmer temperature of 20-25°C (68-77°F) would make them grow faster. Some people in the south plant it in courtyards or living rooms, emitting a faint fragrance. There are restaurants and stalls that plant purple perilla at the entrance, with pots of branches and leaves stacked one after another, looking lovely and lively.

When I was a child, I often went with my classmates to a dapaidang (food stall) called "AhWang" on Haoxian Road after school in search of food. And we ordered a lot of their fried snails with perilla leaf. Ah Wang has a chubby female proprietor. When a customer comes in, she squints her smiling almond-shaped eyes, says "find yourself a seat", and quickly welcomes the newcomer into the stall.

Her son is the only chef in the kitchen, and she herself doubles as the cashier and waitress, acting quickly and efficiently. Among various snacks, the fried snail with purple perilla leaf is often the one that sells out the fastest, because after the snail is brought in, it takes two days to prepare, and cannot be restocked immediately.

Once I saw her put some newly bought snails into a large bucket, and add a few drops of soybean oil into the bucket filled with clean water. She then sat on a small plastic stool using pliers to clamp off the pointed end of the snails, making a "clattering" sound. The fresh snails have a lot of small sand grains to expel, so having them in clean water for two days allows time for them to expel as much sand as possible. Cutting off the tail end of the snail shell is beneficial for allowing seasoning to seep in during cooking and making it easier to also sip out the snail meat when eating.

I still remember that the perilla leaves with fried snails I ate in this store were seasoned with ginger, garlic, fermented soybeans, and chili peppers, and there was cooking wine added to eliminate any fishy smell. The fried snail meat is rich and delicate, with a delicious and refreshing flavor.

The best snails can be found in small streams or paddy fields in the mountains on the outskirts of Guangzhou. The snails found here are more pure because of the clean environment, so there is almost no sand to eliminate and the meat is crispy and juicy, making them a unique treat difficult to find in big cities. Perilla leaves not only add fragrance to snails, but also have the salutary effects of dispelling cold and regulating gas in one's body. The combination of the two is a rare delicacy and also a healthy, low-carbon green food.

During this year's Mid-Autumn Festival, on the night of the full moon, will you stir fry a plate of river snails with perilla leaves and gather with your family to chat and enjoy the beauty of the moonlight?

青瓜卷，让人欲罢不能

在我们日常生活中，青瓜似乎是一种再普通不过的蔬菜，但其实，它所蕴含的营养价值远远超出我们的想象。青瓜不仅富含多种维生素与矿物质，还能够有效增强免疫力，促进新陈代谢，并对内脏器官起到良好的保护作用。更令人称道的是，青瓜富含纤维与水分，但脂肪含量却为零，食用后能迅速带来饱腹感，用青瓜减肥能事半功倍。加上它物美价廉，所以时常被作为最接地气的"平民蔬菜"出现在普通家庭的餐桌上。

青瓜的表皮特别爽脆，因此保留青瓜皮能够最大程度地展现其独特口感，但是可以将较为粗糙的瓜瓤去除。那么，如何能做好一道既保留爽脆外皮又去除瓜瓤，同时味道鲜美的青瓜菜肴呢？

我曾有幸向半岛餐饮集团的创始人、米其林星级餐厅半岛名轩的行政主厨利永周先生请教。他的创意料理青瓜卷，让人能够在尽享爽脆青瓜皮的同时，避免吃到不受欢迎的瓜瓤。他告诉我，传统的凉拌青瓜做法多为将青瓜拍碎，撒盐拌蒜，直接生吃。然而，在拍青瓜的过程中，青瓜因被拍碎而容易失去大量汁液，口感略显逊色。为此，他专门研制出一道青瓜卷菜式，完美地保留了青瓜外皮，口感极佳。

我曾经到半岛名轩观摩学习，发现做好此菜的关键在于刀工。厨师使用1把极其锋利的刀，先从青瓜表面切出1个小口，左手稳稳按住青瓜，右手则顺着切口切下去约1厘米，开始向顺时针方向旋转刀刃，直至将整片青瓜皮连带约0.5厘米的嫩肉完美切下。切好的青瓜皮浸泡在白醋调制的酱汁中以后，入口清爽嫩脆、汁水充盈。那是一种与吃凉拌青瓜完全不一样的舌尖体验。做好的青瓜卷，被置于颜色柔和的光亮釉面瓷盘上，与碟子交相辉映，相得益彰。

当然，这样精湛的刀工并非一朝一夕可得，但想在家中制作青瓜卷亦非难事。你可以用片刀将清洗干净的青瓜横切成片，去掉其中的瓜瓤，将整条青瓜片浸泡在醋中约1小时，捞起来卷成卷状，精心摆盘。这样做出来的青瓜卷虽不及大师之作，但仍别有一番风味哦。

简约的食材，经过巧妙的处理，往往能够彰显出真正的匠心。或许，当年米其林评审员来到半岛名轩时，就品尝过这道外表朴实、口感独特、脆爽可口的青瓜卷呢。这道菜既展现了简约之美，亦蕴含了精致的风韵，正可谓，平凡中见臻至。

青瓜卷
Cucumber Rolls

Simplicity is Key, What Makes This Cucumber Roll So Irresistible?

The cucumber might seem at first glance to be quite an ordinary vegetable, yet it has substantial nutritional value that bolsters our immune system, fosters healthy metabolism, and provides effective protection for our internal organs. At the same time, cucumbers are rich in fiber and moisture, while boasting zero fat content. Eating cucumbers can help satisfy the appetite, making them an excellent choice for those pursuing weight loss goals. Moreover, owing to their mild flavor and budget-friendly price, cucumbers are a favorite on the family dining table.

The skin of the cucumber presents a delightful crunch yet the pulp is a bit rough to eat. Can we craft a dish that features the cucumber's skin, eliminates the pulp, and still tastes great? The answer is a resounding yes!

I once had the privilege of consulting with Mr. Li Yongzhou, the visionary behind the Peninsula Restaurant Group. His Bandao Mingxuan Restaurant has been awarded the prestigious Michelin One Star accolade a number of times. Mr. Li's ingenious creation, his signature cucumber rolls dish, beckons us to savor the sumptuous essence of cucumber skin, elegantly sidestepping the complexities of the seeded interior. During our conversation, he divulged a fascinating insight: our traditional chilled cucumber salad, typically pounded and doused in salt, then intermingled with garlic before being consumed raw, inadvertently loses its juiciness, resulting in a somewhat diminished flavor. With this is mind, he has crafted his remarkable cucumber roll recipe, a culinary gem designed to showcase the cucumber's skin while ensuring a taste that remains intact.

I once had the chance to step into the bustling kitchen of his restaurant, witnessing firsthand the chef's mastery in crafting this dish. What I discovered was that the secret of this dish is tied to the precision of the cutting technique. The chef wields an impeccably sharp blade, initiating the process by making a calculated incision in the cucumber's skin. With his left hand, he secures the cucumber against the cutting board, executing a meticulous cut roughly one centimeter along the initial slice. Thereafter, the cucumber begins to gently rotate in a clockwise motion, guided by his left hand, while the knife follows suit, methodically peeling off the cucumber's skin until only about half a centimeter of flesh remains—a tantalizing marriage of taste and texture. After a soak in a delicate white vinegar concoction, the cucumber skin emerges as a delightful balance of sweetness and tanginess, offering a perfect remedy for the balmy summer months. A single bite combines crispness, juiciness, and chewiness, an experience distinct from eating the whole cucumber. Presented in the elegant form of rolls, this culinary masterpiece is a visual spectacle that complements the dish, creating aesthetic harmony for the senses.

Mastering the art of knife work to the degree of Chef Li requires tremendous dedication, but you can recreate this dish at home without such knife precision. Begin by meticulously washing the skin of a large cucumber. Proceed to make several uniform slices, setting them aside. Gently remove the cucumber's seeded core and immerse these thin slices into a vinegar-based sauce for approximately an hour. When you take them out, you'll notice the cucumbers have taken on a pleasing softness. Now carefully roll them into spirals and place them upon a plate.

The mastery of humble ingredients is a testament to culinary ability. Perhaps, during the judging of the Michelin panel at the Bandao Mingxuan Restaurant, they too fell in love with this unassuming yet remarkably crisp cucumber roll—an embodiment of both simplicity and sophistication.

我在品尝青瓜卷
I am enjoying the cucumber rolls bite by bite

与半岛餐饮集团创始人、米其林星级厨师利永周先生（左一）在结束采访后合影
A photo taken after an interview with Mr. Li Yongzhou (first from the left), founder of Peninsula Catering Company and a Michelin-starred chef

奥运冠军的潮汕蚝仔烙

潮汕的蚝仔烙以其外焦里嫩、香脆可口而闻名，且营养也极为丰富。这道经典小吃的主料非常简单，仅包括生蚝和鸡蛋两样。

首先，鸡蛋是一种平凡却必不可少的食材。自幼，每天早餐妈妈都会逼着我吃鸡蛋，或煎，或蒸，或炸，或煮，已经吃成了习惯。哪天若不吃，总感觉生活少了点什么。对于我来说，简直就是"一日不吃蛋，浑身不自在"。至于这道蚝仔烙中的生蚝，选用的是"珠蚝"，此"珠蚝"非彼珠蚝，并非指产于法国南部的珍珠蚝品种，而是指生蚝中个头较小的一种，价格仅为普通生蚝的十分之一。广东、福建（厦门）一带盛产珠蚝，又以潮汕饶平的珠蚝最为出名——这里的珠蚝肉质饱满、鲜嫩肥美，被潮汕食家视为制作蚝仔烙的首选。清代闽督李鹤年曾有诗云，"蛎房风味胜江瑶"（生蚝闽南语里称"蛎房"）。又有唐代诗仙李白说："天上地下，牡蛎独尊。"自古以来，生蚝的美味已有诗词佐证，非凡之味，可见一斑。

虽然我十分钟爱蚝仔烙，但市面上的蚝仔烙常常油脂偏多，酥是酥、脆是脆，虽香口但其用油之多让人望而却步——热油没过了鸡蛋糊，让人瞠目结舌。

无独有偶，我有幸结识了2008年北京奥运会运动员村的中餐总厨张来得师傅。张师傅来自潮汕，烧得一手好菜，精通潮汕美食。我从他那里了解到当年奥运会运动员村里的菜竟然也有一道蚝仔烙，并且深受运动员们的喜爱。

我不禁好奇问他："听说运动员们对饮食健康要求极高，通常避免食用质感酥脆的油煎或油炸的食物，这道蚝仔烙是如何做到的？"

张师傅解释说，运动员村的蚝仔烙与市面上的传统做法大不相同，虽然主要食材相似，但烹饪方法却更加注重健康，只用很少的油。而蚝仔烙的表面也不会出现油炸后的焦化或酥脆感，却依然保留了其鲜嫩的风味。我爱吃，但也不想吃得太过油腻，于是向张师傅讨教了这道少油版蚝仔烙的制作秘方。

　　张师傅建议，可以在水产市场或网上购买"珠蚝仔"。将珠蚝清洗干净，沥干水分，不要提前煮熟，而是直接用来烙制以保留其鲜味和嫩滑。准备地瓜粉、葱花和2个鸡蛋。将地瓜粉与水按1∶1.5的比例调和成糊状，再加入生蚝和生鸡蛋搅拌均匀。喜欢的话，可以加入少量生白萝卜丁，以增加蚝仔烙的鲜甜感。接着，用少量油将锅烧热至约70℃时，将油倒掉，再倒入少许新油。这种方法在粤菜中称为"猛镬阴油"（热锅冷油），其目的是只用少量油烹熟食物而不至于出现粘锅的现象。最后，将混合好的糊倒入锅中，翻转两面烙至金黄，再撒上葱花即可。潮汕人吃蚝仔烙时，通常会搭配鱼露和胡椒粉，蘸一点酱汁，味道更加鲜美回甘。

　　据说，当年这道低油烹制的蚝仔烙，不仅深受中国运动员喜爱，还吸引了不少外国运动员。其实，这种低油烹饪法也更加贴近"蚝仔烙"中"烙"的原意——几乎不使用油或只用极少的油，通过器皿直接接触食物传导热量将其煎熟。毕竟，蚝仔烙不是蚝仔炸，用这种低油的方式制作，或许正是对这道传统美味的最佳诠释吧！

潮汕蚝仔烙
Chaoshan-style Oyster Omelet

The Olympic Champion's Chaoshan Oyster Omelets

(Chaoshan-style) Oyster Omelets are crispy, delicious, and nutritious. The main ingredients are oysters and eggs.

The egg is such a simple ingredient that, because of its nutritional value, my mum has coerced me into eating every day for breakfast from childhood, whether fried, steamed, boiled, or slow-cooked. The oysters used are the baby oysters or so called "pearl oysters", not the French pearl oyster species, but those found in Guangdong, Fujian province. In Guangdong, the pearl oysters from Raoping, Chaoshan, are the most famous; evenly sized, round, with a plump texture, and considered the first choice by Chaoshan restaurants for making oyster omelets. Li Henian, the governor of Fujian in the Qing Dynasty, wrote a poem that goes, "The tastes of oysters fare better than that of Jiangyao—a type of shelled fish." Li Bai, a poet of the Tang Dynasty, said that "oysters are respected in Heaven and on the Earth". Oysters are delicious, enjoyed nowadays and also by ancient poets.

The oyster omelet one usually sees on the market is crisp and a little oily. Sometimes when pan fried, the oil will cover up the entire omelet.

I met Mr. Zhang Laide, who served as the chief Chinese chef of the Athletes' Village for the 2008 Beijing Olympic Games. Zhang is from Chaoshan and is proficient in all kinds of Chaoshan cuisine. I learned from him that the oyster omelet dish was on the menu in the 2008 Olympic Athletes' Village.

"Oh really?" I asked him,"I heard that athletes were not supposed to eat crispy fried food? How could this oily oyster omelet dish make it onto the menu?"

It turned out that the oyster omelets served

in the Athletes' Village were not the kind commonly seen in the market. Although the materials used were generally the same, the cooking method was different, and the ones eaten by athletes were much less oily, and there was no crispness or burnt areas on the surface of the food. He shared a recipe that uses much less oil and became my favorite.

He said, "pearl oysters" could be purchased in the aquatic market or online. Wash the oysters, drain the water, and keep them raw for later use. Prepare sweet potato starch, shallots and two eggs. Mix the sweet potato starch with water at a ratio of 1 : 1.5 to make a paste, then add oysters and raw eggs and stir well. If you like, you can also add a small amount of diced raw radish to increase the sweetness of the omelet. Heat the saucepan, pour in a little oil, heat the saucepan to about 70°C(158°F), pour out the hot oil, and then pour in a little new oil. This technique is called "hot pan cold oil" in Cantonese cuisine tradition, and it aims to use only a little oil to cook fresh food without having stick to the pan. Pour in the oyster egg mixture and fry until golden on both sides. In Chaoshan, the way to eat this food is to add a little pepper and dip it in fish sauce.

It is said that in addition to being loved by Chinese athletes, this dish also attracted many fans among the foreign athletes at the Athletes' village. In fact, this very-little-oil cooking method is closer to the original meaning of "Lao" or "baked in a pan", in the name of "蚝仔烙 (hao zai lao) ". In this cooking method, almost no oil or only a small amount of oil is used, to come in contact with the food, conduct heat until it's cooked. After all, "baby oysters baked in a pan" is not fried oysters, and frying does require much more oil. But this minimal oil cooking method is a fresh way to enjoy this popular dish. Using this very-little-oil method may very well be the perfect interpretation of this traditional delicacy.

与2008年北京奥运会运动员村中餐总厨张来得师傅在采访结束后的合影
A photo taken after the interview with Chef Zhang Laide, chief Chinese chef of the 2008 Beijing Olympic Athletes' Village

南做北味韭菜合子

虽说我生于广州，但母亲的家乡在河北。从小到大，我没少吃姥姥亲手制作的韭菜合子。她做的韭菜合子，猪肉肥瘦得当，馅料清香甘甜，没有丝毫腥味。多年过去，我始终珍藏着她的韭菜合子配方。

鸡蛋韭菜合子是中国北方地区，包括东北地区及山东、河北、山西、陕西、北京和天津等地广为流行的经典小吃，在许多地方，它还是节日佳品。俗语道："初一饺子初二面，初三合子往家转。"春节期间，大家习惯走亲访友，说着拜年的祝福。但在大年初三，按照传说，是不宜外出的，因为这一天为"赤狗日"。据说，"赤狗"是熛怒之神，为避开凶神的影响，尽量不要出门，最好是在家里早睡晚起。而北方有个特别的习俗，就是宅在家里做韭菜合子。合子的外形酷似一个金元宝，象征着团团圆圆、和和美美，还有财富滚滚、"把钱往家赚"的意思。在初三吃合子，寄托了人们希望避开"赤狗"带来的不良影响，祈愿来年和气生财、福运绵长。

韭菜合子味道鲜美，且制作食材易于获取，深受家庭厨师们的青睐。制作韭菜合子的食材包括鸡蛋、猪肉、韭菜、姜、粉丝和虾米等，均为日常所见的食材。以下是我珍藏多年的韭菜合子配方。

首先，准备七分瘦三分肥的猪肉、普通面粉、温开水、生姜、海盐、花生油、芝麻油、干虾米、粉丝、鸡蛋以及新鲜韭菜。将普通（非低筋）面粉与温水揉成面团，醒一会儿备用。然后，将猪肉与小块生姜一起放入搅拌机打碎，加入适量海盐、花生油和芝麻油，搅拌均匀。干虾米和粉丝则用沸水泡软，剁碎后加入猪肉馅中。热油起锅，将打散的5个鸡蛋炒至一半生、一半熟，拌到肉馅中。最后，将洗净的韭菜切成小段。韭菜要生煎才好吃，生煎后将它拌到猪肉馅中。最重要的是包合子：将醒好的面团分成小块，包上调好的馅料，捏紧封口。在锅里放少许油，放入韭菜合子，两面煎至金黄后，改文火盖上锅盖烘熟就行了。

　　这样用文火煎出来的韭菜合子，外形宛如金元宝，色泽诱人，馅香扑鼻，人见人爱。

南做北味韭菜合子
Southern-Made Chive Boxes with a Northern Flavor

Northern Chinese Egg and Chive Boxes Made the Cantonese Way

Although I was born in Guangzhou, my mom's hometown is Hebei in Northern China. From childhood to adulthood, I ate a lot of my Grandma's pan-fried Chinese egg and chive boxes. Her dish uses pork that is just right, not too lean and not too much fat, fragrant and delicious. Over the years, I have kept her recipe.

Chinese egg and chive boxes are a very popular traditional snack in Shandong, Hebei, Shanxi, Shaanxi, Beijing, Tianjin and other regions in Northern China. It is also a festival food in many places. As the saying goes, "Eat dumplings on the first day of the Chinese New Year, noodles on the second day, and on the third day, bring chive boxes home." This is because every Spring Festival, we are accustomed to visiting relatives and friends, and exchanging New Year's greetings. However, on the third day of the Chinese New Year, it is not advisable to go out because this day is the Red Dog Day.

In legend, "Red Dog" is the "God of Wrath", and one should go to bed early and wake up late and stay at home to avoid running into the "Red Dog". The custom in Northern China is to stay at home and make egg and chive boxes. First, the box represents reunion, harmony, and beauty. Also, the box looks very much like a Chinese gold ingot, so shaping the boxes at home also implies "bringing money home". So, staying at home on the third day of the Chinese New Year and making egg and chive boxes could represent avoiding the negative impact of the "Red Dog" and also being able to generate wealth and increase financial resources in the coming year.

The ingredients for this egg and chive box are easy to obtain, which is helpful for home cooking. The ingredients include eggs, pork, leeks, ginger, vermicelli and shrimp, all of which are commonly used. The following is my Grandma's recipe. Obtain pork that is 70% lean and 30% fat, flour, warm water, a piece of ginger, sea salt, peanut oil, sesame oil, dried shrimp, vermicelli, eggs and chives.

First, use regular (not low gluten) flour and warm water to knead into a dough and set it aside for later use. Mix the pork and a piece of ginger in a blender, then add sea salt, peanut oil, and sesame oil to blend thoroughly. Soak the dried shrimp and vermicelli together in boiling water until soft, chop them on a cutting board with a knife into mince, and then put them into the blended pork filling. Heat oil in a pan and stir-fry 5 scrambled eggs until they are half cooked; then add the pork filling. Wash the chives thoroughly and cut them into small pieces. The chives should be fried raw for the best flavor, then add them to the pork filling. The most important thing is how you wrap the box. Take out a piece of dough and knead it, roll it out, put some filling in, and wrap it well. After adding a little oil to the pan, fry the box on both sides and then cover it with a lid, simmering on low heat until cooked.

The box cooked on gentle heat like this really looks like a gold ingot, and everyone loves it.

粤式水果蔓越莓捞桃胶

近年来，广东的水果大量出口至世界各地，我在国外拍摄节目时就曾见到过许多如"菠萝的海，香飘四海"（Pineapple from Guangdong, China, Fragrant Everywhere)，又或者是"广东喊全球吃荔枝"（Guangdong is Calling on Everyone to Try Lychee)等有趣的大型广告标语。然而，这些现象是近几年的事。早年我在海外求学时，若想吃到广东的水果，真是难如登天。加上我当时身处大半时间都气温低下的美国芝加哥，能碰到新鲜水果都觉着新奇。幸好，我喜欢上了当地一款特产水果干，名叫蔓越莓。这种水果主要产自美国马萨诸塞州、加拿大魁北克省以及智利。蔓越莓与我们的山楂有几分相似，酸酸甜甜的，总让我回忆起小时候吃冰糖葫芦的情景。作为水果干，蔓越莓口感独特，酸甜交织，热量低且富含矿物质，便于随身携带和保存。于是，我常备一小袋在身上，偶尔拿出来解馋。

历史记载，蔓越莓在早期的北美属于珍稀食品。1677年，北美殖民者曾将它作为贡品献给英国国王，那一次随行的贡品还有2大桶印第安玉米和3 000条鳕鱼。经过漫长的航程，鳕鱼和玉米已经腐烂，唯独蔓越莓依然新鲜如初，并以其原貌出现在了查理二世面前，这归功于其富含天然防腐剂和维生素C。从此，蔓越莓名声大振，并在殖民时期成为新英格兰水手远航时必备的食物。

回国后，我在家乡广州工作和生活。广州素有"水果之乡，美食之都"之称，可吃到的美味太多了，尤其是各种新鲜水果，于是我再不用吃水果干了，但仍会偶尔怀念蔓越莓的滋味，因为那是我在异乡时的一丝慰藉。在一次朋友的聚会上，我再次遇见了它。好友将蔓越莓与桃胶搭配，并铺上了一层新鲜的广东特色水果丁——龙眼、荔枝以及火龙果。品尝后，我惊喜地发现，桃胶与蔓越莓干，以及各种新鲜水果丁搭配，味道竟然相当不错。

桃胶原产自中国云南，是桃树上分泌的胶状物，其凝结成块，形似橡皮糖，口感Q弹，但并不甜腻。桃胶的价格相对亲民，获取也很方便。如果想减轻桃胶的寒性，可以在烹煮时加入几片姜片。

这道粤式水果蔓越莓捞桃胶，清凉可口，是夏日消暑的佳品。制作时，可提前1天将桃胶泡发，挑去杂质，沥干水备用。因桃胶性寒，煮时可在压力锅中加入姜片与适量清水，再加一些红糖，盖好锅盖，压煮约15分钟。煮好后把姜片取出弃之，将桃胶放凉后放入冰箱冷藏。食用时，将自己喜欢的水果切丁，与蔓越莓干一起撒在冰镇过的桃胶上。喜欢的话，再淋上一些椰浆，便大功告成！

这款中西合璧的甜品将来自美国的蔓越莓与中国西南的桃胶巧妙结合，你又会选用哪些广东特色水果来与它们搭配呢？

水果蔓越莓捞桃胶
Peach Gum with Mixed Fruits and Cranberries

Cantonese Fruits and Cranberries with Peach Gum Dessert

In recent years, Guangdong fruits have been exported in large quantities to different countries and regions around the world. I have come across interesting advertising slogans such as "Pineapple from Guangdong, China, Fragrant Everywhere" or "Guangdong is Calling on Everyone to Try Lychee" while travelling abroad. However, this was not the case in the early 2000s. When I was studying in the US, particularly in Chicago, not only was it difficult to find Guangdong fruits, but even finding fresh fruits overall was challenging due to the long-time cold weather. Nonetheless, I quickly fell in love with a local specialty fruit called cranberry. Cranberries are primarily grown in Massachusetts in the United States, Quebec in Canada, and Chile. This berry bears some resemblance to our hawthorn berry from Guangdong, which reminds me of the childhood anecdote of eating " 冰糖葫芦 " (candied haws on a stick). Cranberries have a delightful combination of sweetness with a

hint of sourness. Moreover, when they are dried, their flavor intensifies, making them even more enjoyable. Consequently, I would often carry a small bag of dried cranberries with me.

According to historical records, cranberries were a rare food in the early 17th century from North America and were once presented as a gift to the British king by the North American colonies in 1677. The gifts at that time also included two large barrels of Indian corn and three thousand sea cod. Unfortunately, after a long voyage, both the cod and corn rotted, and only cranberries survived and appeared in their fresh state before King Charles II, largely due to the abundant natural preservatives and vitamin C in them. From then on, cranberries gained quite a reputation, becoming a staple food for New England sailors during the colonial period.

Later, I returned to China, and I would still

miss cranberries from time to time because they reminded me of the familiar taste that comforted me and helped ease my nostalgia while in a foreign country. I had the pleasure of rediscovering them at a friend's party, where they were used as a key element in a delicious dessert. My friend combined them with Chinese peach gum and layered the dessert with fresh Guangdong fruit cubes—longan, lychee, and dragon fruit on top, finishing it off with a touch of coconut milk. The resulting taste was light, smooth, silky, and provided sheer enjoyment!

The origin of this peach gum in China is the Yunnan Province in the Southwest of the country. It is made from flowing sap of peach trees, condensed into little cubes bound on the peach tree stems. It tastes a bit like gummy bears, but it's not overly sweet. The price of peach gum in the market is not high, making it easily accessible. Many people like to add ginger during the cooking of peach gum to make it easier for the body to digest.

In the summer, preparing a refreshing and delicious treat of peach gum with mixed fruits and cranberries can help reduce inflammation in the body. Begin by soaking the peach gum a day in advance, ensuring to remove any impurities after soaking, and then drain the water for later use. Since peach gum is considered to have a "cold nature" in Chinese herbal medicine, you can add two pieces of ginger when placing it into the pressure cooker along with some water in order to add "heat", which also makes it easier to digest. Next, add brown sugar, cover, and press for about 15 minutes. Afterward, discard the ginger slices, pick out the peach gum, let it cool, and refrigerate it until chilled. When ready to serve, dice your favorite fruits and place them alongside dried cranberries on the chilled peach gum. For an extra touch, add some coconut milk. Now, you're all set to enjoy a delightful summer treat!

This delectable fusion of Chinese and Western foods brings together dried cranberries from the northeastern part of the United States and peach gum from the southwestern part of China. Now, the question is, what Guangdong fruit would you choose to pair with these delightful ingredients if you had the opportunity to try or make this dessert?

爱情草烤羊肉——幸福的味道你闻到了吗？

因为做美食节目，我结识了不少意大利厨师，随着大家的友谊加深，我们时常聚在一起畅谈美食与生活。在与他们的交流中，我发现，他们很喜欢在做意大利菜时加入一款在欧美地区被称为"爱情草"的香料——迷迭香。无论是煎焗羊排、牛排，还是烤蔬菜、烤土豆，乃至圣诞大餐或复活节盛宴，迷迭香都能让菜肴散发出独特的芬芳。

迷迭香堪称"百草之王"，在中国的广州、上海、北京等地广泛种植。作为一种常绿灌木，它在夏天开出如海洋露珠般晶莹的小蓝花，宛如大自然的精灵。在欧美，人们用它编织成节日花环，悬挂于门前、房间或户外，散发出阵阵草木清香。它带有松木韵味，浓郁而略带刺激，对人体有百益而无一害。因为它散发出来的气息迷人又特殊，欧美人常将它与浪漫的爱情联系在一起，称之为"爱情草"。英国一首古老的民谣中写道："当你爱上一个人时，带着你的思念，走进开满迷迭香的田野诉说心事，很快，心上人就会出现在你面前。"

迷迭香自古以来便被视为智慧与力量的象征。古希腊的哲学家们常在头上佩戴用迷迭香编织的花环，以增强记忆力。时至今日，希腊的学生在考试前仍会在家中燃烧迷迭香，期盼它的香气能激发灵感与动力。

我曾与一位在广州四季酒店担任主厨的意大利厨师菲利波谈论过迷迭香。他认为广州的迷迭香与意大利的相差无几。尤其与羊肉搭配时，迷迭香能将羊肉的风味提升至极。他向我传授了一道迷迭香烤羊腰肉配蔬菜（lamb loin with rosemary and assorted vegetables）的秘方。其关键在于取材，要用羊腰肉，这是羊背部最后一根肋骨与臀部之间位置的那块肉，肉质细嫩多汁，脂肪含量极低。先将羊腰肉撒上海盐、黑胡椒粉、橄榄油和迷迭香粉，腌制20分钟。在一口干净的煎锅中放入黄油和一株新鲜迷迭香，再放入羊腰肉慢煎，煎的时候将黄油不断淋在羊肉上，直到表面有点焦黄。接着将煎好的羊肉包入锡纸，并在原本有羊皮的一侧埋入一枝迷迭香，放入预热至200℃的烤箱，烤制约10分钟即可。若喜爱浓郁点的风味，还可以用黄油煎些洋葱丝，作为配菜与羊肉同食。

在忙碌了一整天后，回到家中，吃着黄油迷迭香煎烤羊肉，甜蜜与幸福来得如此简单！

迷迭香烤小羊排
Rosemary Roasted Lamb Chops

Roasted Lamb with Rosemary—Do You Smell Love in the Air?

I've met many Italian chefs through participating in cuisine and culture activities. After getting to know each other, we often gather and chat. During our interactions, I discovered that they really enjoy using a European and American spice commonly known as rosemary in their Italian dishes. Whether it's fried lamb chops, steak, mixed vegetables, roasted potatoes, or festive occasions like Christmas and Easter feasts, rosemary is a staple in high-end Western restaurants and important Western festivals.

Rosemary can truly be considered the king of all herbs. In China, rosemary is primarily produced in major cities such as Guangzhou, Shanghai, and Beijing. It belongs to the evergreen shrub family, blooming with small blue flowers in the summer, resembling tiny, crystal-clear dewdrops in the ocean. In Europe and America, people use it to weave various festive wreaths as decorations, hanging them on doors, indoors, and outdoors. Its fragrance carries hints of grass and pine, with a strong and slightly stimulating scent that is beneficial to the human body without any harm. Due to its enchanting and distinctive aroma, in Western countries, it is often associated with romantic love and referred to as the "love herb". An ancient English ballad states, "When you fall in love, bring your thoughts to a field full of rosemary to express your feelings. Soon, your beloved will appear before you."

Since ancient times, Greek philosophers often wore wreaths made of rosemary on their heads to help improve their memory during exams. Even today, Greek students often burn rosemary at home before exams, believing its fragrance can inspire and uplift them.

I once had a conversation about rosemary with an Italian chef named Filippo, who served as the head chef of the Italian restaurant at the Four Seasons Hotel in Guangzhou. He mentioned that the taste of rosemary produced in Guangzhou is comparable to that in Italy. According to him, it pairs most harmoniously with lamb. He shared the secret to preparing a delightful dish—lamb loin with rosemary and assorted vegetables. The key lies in selecting lamb loin, the meat between the last rib on the sheep's back and the buttocks. This meat is smooth, tender, juicy, and low in fat. Start by seasoning the lamb loin with sea salt, black pepper, olive oil, and rosemary powder, then marinate for 20 minutes. In a clean frying pan, add butter and rosemary stems, followed by searing the lamb. While searing, continuously pour butter over the lamb until the surface turns slightly golden brown. Transfer the lamb onto a sheet of aluminum foil, bury a rosemary plant on the side originally covered with skin, wrap it up, and place it in the oven. Set the temperature to around 200°C(392°F) and bake for about 10 minutes. If desired, you can also cut some onions into shreds, fry them in butter in a pan, and serve them as a side dish with the lamb.

On a night after a busy day, returning home to creamy rosemary-seared and roasted lamb makes sweetness and happiness so simple!

用广东的食材烹饪著名的西非菜肴

如果要用一种颜色来描绘西非大陆，我会选择土红色——那是一种原始又奔放、炽热且充满生命力的颜色。如果要用一道菜来代表非洲，我会选择土红色的乔洛夫炒饭（Jollof Rice）。这道菜家常且风味独特。对于许多人来说，非洲大陆遥远、神秘，而非洲的美食对许多人来说更是知之甚少，但乔洛夫炒饭却是非洲众多美食中已经享誉全球的一款。它的主要食材包括大米、番茄、大红辣椒、小米椒和洋葱等，在广东，我们也可以轻松找到这些食材，制作这道经典的非洲美味。

乔洛夫炒饭的起源可以追溯到西非塞内加尔境内的沃洛夫民族。其名字"Jollof"源自沃洛夫语中的"Wolof"，也就是沃洛夫人的意思。后来，这道菜随着沃洛夫人的脚步流传至尼日利亚、马里、加纳等西非国家，成为当地人餐桌上的常见美食。

我第一次吃乔洛夫炒饭，是应邀前往尼日利亚前驻穗总领事的家中做客。当时，总领事和他的夫人亲自迎接我们。那天，总领事夫人身着鲜红华丽的民族服饰，笑容满面地将我们迎进客厅，为我们端来茶水。尽管他们家有从尼日利亚随同而来的专业厨师和侍应，负责照料他们的日常饮食，但为了表达对客人的热情款待，夫人系上围裙、亲自下厨为我们炮制这道大餐。我不禁感慨夫人的淳朴特质以及发自内心的好客。这也是真正意义上的友好与热情，有别于场面上的客套与寒暄。

厨房里，所有食材早已准备妥当。尼日利亚菜肴中，番茄、洋葱和棕榈油几乎是不可或缺的元素，乔洛夫炒饭自然也少不了它们。传统的尼日利亚乔洛夫炒饭的主要食材包括长粒米、番茄、大红辣椒、伯纳特辣椒（非洲特有的一种辛辣辣椒）、棕榈油和洋葱。调味料则是鸡精、盐、黄色咖喱粉、百里香、蒜粉和2片月桂叶。并且，非洲人偏爱鲜艳的色彩，因此也会在炒饭中加入红艳浓郁的番茄酱，这也是乔洛夫炒饭的点睛之笔。

夫人娴熟地将大红辣椒、番茄、洋葱和伯纳特辣椒放入搅拌机搅成泥状，接着倒入小锅中煮至水分挥发，备用；然后，取一口煎锅，倒入棕榈油与洋葱粒翻炒5～8分钟，加入番茄酱继续翻炒约15分钟，再倒入刚刚煮好的辣椒番茄泥，慢火炒5分钟后，加入所有剩余的香料；最后将洗净的生米倒入锅中煮熟，盖上锅盖并用锡纸包裹后再放入烤炉焗20分钟。一道色香味俱全的乔洛夫炒饭就出炉了！

当我吃下第一口时，浓郁的番茄香味与辛辣的辣椒味完美交织，仿佛将我带到了非洲辽阔的大草原，那鲜美浓郁的滋味让人不禁闭上眼睛、慢慢品味、久久难忘。我和总领事一家围坐在餐桌旁，听着夫人独具特色的、极其爽朗又慢悠悠的笑声，感受到她成功准备了一桌子佳肴的喜悦与满足。其中，乔洛夫炒饭的简单制作过程，配上经济实惠的食材，造就了绿色、低碳的一流料理。

在尼日利亚，还有一种耗材比这更少、经济实惠版的乔洛夫炒饭，被称为"调制炒饭"（concoction rice），仅需大米、大红辣椒和番茄，便能成就一番风味。

尼日利亚乔洛夫炒饭
Nigerian Jollof Rice

Making a Well-Known West African Dish Using Cantonese Ingredients

If I were asked to depict the West African subcontinent with one color, I would use earth red. Primitive yet passionate. If I were asked to choose a food to represent Africa, I would choose the earth-red colored Jollof Rice, which is friendly and distinctive. For many people, the African continent is distant and mysterious, and when it comes to the cuisine from this continent, what we know are few and far between. With that said, Jollof Rice is probably one of the most famous African cuisines in the world. Its main ingredients include rice, tomato, red bell pepper, millet pepper and onion. It can also be easily cooked in Guangdong with local ingredients.

Jollof Rice originated from the Wolof nationality in Senegal in West Africa. The name "Jollof " of this dish comes from the Wolof language, which means Wolofer. Wolofer is one of the ethnic groups in West Africa, and later this dish spread to other West African countries including Nigeria, Mali, Ghana, and so on.

The first time I ate Jollof Rice was when I was invited to the home of a former Nigerian Consul General in Guangzhou. On that day, the Consul General personally welcomed us with his wife who was dressed in bright red and splendid ethnic costumes. In fact, the Nigerian Consul General brought authentic Nigerian chefs and helpers from Nigeria, who were usually responsible for taking care of the food and daily life of the host family. However, in order to express a warm welcome to the guests, on that day, the wife of the Consul General took on the role of master chef, tied on an apron, and personally cooked a homemade meal for us. I couldn't help but feel immense joy because of their warm hospitality and simple nature. This is also a true sense of friendship and enthusiasm, different from the petty politeness on the surface.

I followed her into the kitchen to see her prepare the dish and saw that various ingredients were ready on the kitchen table. The most commonly used elements in Nigerian cuisine are probably red tomatoes, onions, and palm oil, which can be seen in almost every dish. And it is naturally indispensable in Jollof Rice. The main ingredients of traditional Nigerian Jollof Rice include long grain rice, tomato, red bell pepper, scotch bonnet, palm oil and onion. Seasonings include chicken essence, salt, yellow curry powder, thyme, garlic powder, and two pieces of bay leaves. In addition, because Africans like bright colors, as can be seen in the outfits or food they use, in order to add color to the food they eat such as Jollof Rice, they would add red concentrated tomato paste (purée).

Firstly, put the red bell peppers, tomatoes, onions, and scotch bonnet into a blender and stir until they are smooth. Pour them into a small pot and cook until the water evaporates. Set aside for later use. Take another frying pan and stir fry palm oil and chopped onions for 5-8 minutes. Then add concentrated tomato sauce and stir fry for about 15 minutes. Pour in the previously made mashed sauce. After frying for

5 minutes, add all the seasoning and pour the washed raw rice into it to cook. Cover the pot with tin foil and bake for 20 minutes to eat.

When you taste this Jollof Rice, it has a strong tomato flavor, and is the real deal—it's got the looks, the aroma, and the flavor to knock your socks off! It's a delightful combination of salty, sweet and a refreshing aroma that allows you to close your eyes and imagine the African savannah. This visit made me feel the approachability, freedom, and openness of African culture. When we sat together, the Consul General's wife's characteristic hearty and lingering slow laughter put us at ease and showed her joy and satisfaction from having made something so beautiful. Economical ingredients can be stirred in a blender to create delicious sauces, coupled with simple preparation methods and rice; a green, low-carbon and first-class dish can be created!

In Nigeria, there is also an economical version of Jollof Rice with even fewer ingredients, called "concoction rice". The ingredients include only rice, red bell peppers and tomatoes.

生姜柠檬水

立秋时节的到来，标志着广东的天气逐渐转凉。古人视立秋为夏秋之交的重要节点，极为重视这个节气。虽然立秋后仍会热上一段时间，但此时正是开始注重祛寒保暖的时节，尤其适合食用一些温补的食材，而生姜无疑是最佳选择之一。《中国药典》记载："（生姜）解表散寒，温中止呕，化痰止咳。"除了温补，生姜还能有效化痰，预防秋季常见的咳嗽与感冒。

你知道吗？生姜与柠檬堪称天作之合。

几年前，我家楼下院子里忽然冒出一棵"飞来树"。起初，这棵小树并不起眼，低矮且瘦弱，枝干纤细，叶子稀疏，根本没有引起我的注意。每年到了立秋，也就结出几颗零星的青柠檬，三三两两、不成气候。然而，前年邻居砍掉了旁边高大的椰子树之后，这棵柠檬树终于得到了充足的阳光照射，竟然枝叶繁茂起来。春天时，花香四溢，引得蜜蜂纷至沓来。到了立秋，小青柠檬如同肥皂泡般一颗颗冒了出来，数量竟然是往年的10倍。柠檬成熟后变成金黄色，果实饱满，砰砰地往地上掉。

某天早晨，我瞧见地上滚落了好几颗柠檬，全是纯天然、无添加的有机种植的果实，弃之不用实在可惜。我记起曾有一次去朋友家做客，看到她用生姜和柠檬煮水饮用，当时我试了一口，感觉味道非常清爽。于是，我决定物尽其用，早晨用柠檬和生姜煮水，或者与水果一起打成汁，喝后顿觉神清气爽，整日心情舒畅。

制作生姜柠檬水时，生姜和柠檬都不需要去皮，因为二者都有药用价值。只需将生姜洗净，和切片的柠檬一同放入锅中，用热水煮沸。熄火后，根据个人喜好加入一些薄荷叶或蜂蜜，浸泡约15分钟即可饮用。如果需要加热，只需轻微加热，无须再煮沸。

　　需要特别提醒的是："早上吃姜，胜似喝参汤；晚上吃姜，等于吃砒霜。"生姜味辛、性温，富含姜辣素等活性成分，适量食用能健脾温胃，加快血液循环，使身体暖和起来。但晚上食用生姜，姜中的活性成分可能会刺激肠道蠕动，影响睡眠。因此，这款生姜柠檬水还是在早上喝为佳哦。

院子里的 "飞来树"
A tree that grew in my yard uninvited, yet turned out to be fruitful

生姜柠檬水
Lemon Ginger Drink

Lemon Ginger Drink

The arrival of Beginning of Autumn signals that the weather in Guangdong starts to cool down gradually. The ancients considered the Beginning of Autumn an important moment marking the transition from summer to autumn, attaching great importance to this solar term. Although it may still be hot for a while after the Beginning of Autumn, it is time to start thinking about dispelling cold and keeping warm, and consuming some warm and nourishing foods. This is an excellent time to consume ginger. According to the *Chinese Pharmacopoeia*, ginger "relieves external heat and dispels cold, warms up to stop vomiting, dissipates phlegm and stops coughing". In addition to its warming properties, ginger can effectively help in dissipating phlegm and preventing coughs and colds in autumn.

And did you know that ginger and lemon are a perfect match?

A few years ago, a tree suddenly appeared in my courtyard. At first, I didn't take it seriously. After all, it hadn't been carefully planted and looked unimpressive, displaying a fragile appearance— not upright, slightly bent, with slender branches and sparse leaves. Every year during the Beginning of Autumn, it would only yield a few lemon fruits, in twos and threes, hardly making an impact. However, just two years ago, after the neighbor cut down the tall coconut tree next to it, the sunlight immediately poured in, and the lemon tree received ample sunlight. Suddenly, its branches and leaves flourished, filling the air with the fragrance of flowers in spring, attracting bees. By the Beginning of Autumn, about ten times as many small green lemons appeared, resembling soap bubbles. When the fruit ripened and turned yellow, it fell to the ground with a bang.

One morning, I approached and saw several

lemons rolling on the ground. They were pure, natural, additive-free, and organically grown fruits. It was a shame not to eat them. I remembered once, during a visit to a friend's house, I saw her boiling water with lemons and ginger. Especially in summer and early autumn, drinking ginger lemon honey water can help cleanse the intestines and stomach, as well as replenish qi and blood. so, I decided to use all these lemons, making ginger lemon water in the morning or blending it with fruits to drink. After consuming it, I felt refreshed, and my mood was pleasant throughout the day.

This autumn health drink is super easy to make, and here are the steps. Ginger and lemon both have medicinal values, and there's no need to peel the skin. Just wash the ginger well. Slice the ginger and lemon and put them together in a pot. Bring the water to a boil, then turn off the heat. You can add mint leaves or honey according to personal taste. Let it steep on the side for 15 minutes, and then it's ready to drink. Strain and drink at room temperature or reheat.

However, I would like to remind everyone that "eating ginger in the morning is better than having ginger soup at night, which is equivalent to eating arsenic". Ginger has a pungent and warm taste, containing active ingredients such as gingerol. Eating some ginger can strengthen the spleen, warm the stomach, accelerate blood circulation, stimulate nerves, and make the whole body warm. However, at night, the gingerol in ginger can stimulate intestinal peristalsis, affecting sleep. So, it's best to drink it in the morning.

波兰国汤——甜菜头汤

甜菜头有其独特的红色，这种色泽源自它富含的花青素。花青素不仅赋予甜菜头鲜艳的颜色，还有助于补血，并能够调节人体肾脏和肝脏的功能。

在波兰，甜菜头被广泛用于制作一种名为"Barszcz"的传统汤品，人们称之为"波兰国汤"。这道汤的原名为"Borscht"，其起源可以追溯到16世纪的东欧。最初，"Borscht"这个词源自斯拉夫语中的"猪草"一词，"猪草"本是该汤原始配方中的主要材料。煮了发酵后的猪草茎叶和花朵以后，这种汤会带有酸味，口感并不出众。直到16世纪末，受到罗马饮食文化的影响，乌克兰人开始将甜菜头加入汤中，这为汤增添了甜味和绚丽的红色，也让这道汤变得更加美味。随后，这道改良后的甜菜头汤逐渐传播到包括波兰在内的西欧和东欧各国，成为庆祝活动和家庭聚会中的经典美食。比方说，在圣诞节前夜的庆祝中，波兰家庭必不可少的12道传统素食中必然包括了这道甜菜头汤。

有一次，我在受邀前往波兰驻穗领事馆举办的国庆招待晚宴上品尝到了这道波兰传统甜菜头汤。那碗清香扑鼻、色泽艳丽的甜菜头汤被盛在精致的波兰手工陶艺碗中，红色的汤水在灯光下闪烁着诱人的光芒，让人不禁食欲大振，迫不及待想要舀一勺送进嘴里。

在晚餐中，我向一位波兰外交官请教了如何制作地道的波兰甜菜头汤。她告诉我，制作4人份甜菜头汤的基本材料包括1千克甜菜头、1根胡萝卜、1个洋葱、少许欧芹根、1.5升水、2瓣新鲜大蒜和1朵蘑菇（或可用干香菇替代），调料包括3片香叶、1小勺白醋、1小勺茴香籽和适量甘牛至、多香果。其中，甜菜汤中的蘑菇最好选用在森林里采摘的野生蘑菇。她还提到说，在波兰不少家庭有去森林采蘑菇的习惯。我好奇地问："难道不怕采到有毒的蘑菇吗？"她笑着说："在波兰，大多数人都有分辨蘑菇哪些不能吃、哪些能吃的经验。如果不确定，可以带到药房，请药剂师帮忙识别。""靠山吃山，靠水吃水，不时不食"，我想，这正是接受大自然馈赠的最好方法。

制作甜菜头汤的过程非常简单：把所有蔬菜去皮，切成约2厘米的方块，放入锅中与调料、水一起煮沸后关火，静置6～8小时。食用时，如果想让汤的色泽更为鲜艳，可以将新鲜的甜菜头切成小块，放入滤网，再将汤舀进滤网里过滤入碗。

这位女士回忆说，好多年前她在上海工作时，想吃地道的甜菜头汤必须去北京购买甜菜，当时的上海几乎找不到这种食材。有一年临近圣诞节，她带着拉杆箱从北京装了满满一箱子的甜菜头回上海，分给同事们做汤。如今，随着中国电商平台的发展，甜菜头在网上商城也能轻松购买到，在广东制作正宗的波兰甜菜头汤也不是难事了。

波兰甜菜头汤
Polish Beet Soup

Beet Soup—The Polish National Soup

Beetroots have a special beautiful red color due to their high anthocyanin content. Anthocyanins have many benefits, such as nourishing the blood and effectively regulating kidney and liver function.

In Poland, local people like to use beetroots to make a soup called "Barszcz", widely considered the "Polish national soup", and with the original name "Borscht", which originated in Eastern Europe in the 16th century. The name "Borscht" comes from an old Slavic word for "Hogweed", which was the main ingredient in the original recipe. That version of the soup had a sour taste because it used fermented hogweed stems, leaves, and flowers. The transformation of borscht began later in the 16th century, when Ukrainians, influenced by the Romans, began incorporating beets into the recipe. Beets were not only affordable and nutritious but also added a sweet taste and a vibrant color to the soup.

This new version of borscht quickly spread to neighboring nations including Poland, and eventually became a cornerstone of Polish local cuisine. Passed down from generation to generation, the soup takes center stage during major celebrations and family gatherings. During Christmas Eve celebrations, Barszcz takes a prominent position among the 12 traditional Polish vegan dishes.

Previously, I had the privilege to attend the Polish National Day Reception in Guangzhou upon the invitation of a Polish diplomat friend. I tasted Barszcz during the meal. The clear and fragrant beetroot soup was served in a handmade Polish ceramic soup bowl, with a bright red and enticing color that warmed my appetite. From the moment I saw and smelled it, I had an urge to scoop a spoonful into my mouth.

I asked her how to cook authentic Polish

Barszcz. She said that the most basic ingredients (enough for four people) include: 1 kilogram of beetroot, 1 carrot, 1 onion, a little of the white part of parsley, 1.5 liters of water, 2 cloves of fresh garlic, 1 wild mushroom (or dried shiitake mushroom can be used); The seasoning and spices include 3 bay leaves, 1 tablespoon of white vinegar, some dill, marjoram and allspice. She also mentioned that the mushrooms in the beet soup should be wild mushrooms from in the forest, and shared that many families in Poland have the habit of going to the forest to pick mushrooms and then take them home to wash, dry and use for cooking. I asked her if she was worried about picking poisonous mushrooms, and she said that in Poland, most people can differentiate between edible and poisonous mushrooms. In Poland, it's also possible to send the collected mushrooms to a local pharmacy for pharmacists to check. There is a Chinese saying that goes "Mountain dwellers live off the land, while coastal folks thrive on the sea's bounty, and stick to what's in season". The use of these natural mushrooms is really a good way to embrace the gifts of nature.

The cooking process of Barszcz is simple. Peel off the skin of all the vegetables, cut them into cubes with a diameter of about 2 centimeters, and mix them together. Heat all the ingredients, spices, and water in a large pot until boiling and then turn off the heat and let it sit for 6-8 hours. If you want the soup to have a brighter color, cut the fresh raw beets into very small pieces and put them through a filter.

This diplomat friend said a long time ago, when she was working in Shanghai, she needed to go to Beijing to buy beets because they were almost impossible to find at any other place in China at that time. One year near Christmas, she went to Beijing, loaded up a full suitcase of beets, and brought them back to Shanghai to share with her colleagues to make beet soup. "However, beets can be bought online these days in China, so that making authentic Barszcz in Guangdong is now comparatively very easy." She said.

巴基斯坦比亚尼香饭

      毫不夸张地说，比亚尼香饭已经成为巴基斯坦美食文化的象征，其使用的香料与食材，正是巴基斯坦饮食精髓的浓缩。比亚尼香饭融合了10种以上当地香料，带来了多重香气的碰撞和味蕾的绝妙体验。这道菜不仅是对巴基斯坦食物的展现，更讲述了南亚数百年历史的饮食文化故事。

      比亚尼香饭的起源可追溯到南亚的莫卧儿王朝（1526—1858年统治南亚大部分地区）。"比亚尼"（Biryani）一词源自波斯语中的"Birian"，意为"烹饪前的油炸"，这是莫卧儿王朝的皇家厨师为每一粒米饭注入浓郁香气的烹饪技法。一些历史学家认为，比亚尼香饭的发明归功于莫卧儿帝国的皇帝沙·贾汗的妻子蒙泰姬·玛哈尔（即泰姬陵的灵感来源）。传说她曾注意到军队士兵营养不良，便下令用丰富的香料、肉类和米饭制作了一道营养丰富的菜肴，这便是比亚尼香饭的雏形。从此，比亚尼香饭在巴基斯坦、印度和斯里兰卡等南亚次大陆地区流传开来。如今，在巴基斯坦，比亚尼香饭不仅是日常饮食的一部分，更是婚礼、开斋节庆祝以及家庭聚会等重大场合的必备主菜。

      一次，我有幸受邀到巴基斯坦驻穗总领事萨达尔·穆罕默德的家中做客。那天，总领事夫人莎夫卡身穿传统服装，为我们准备了一桌精美佳肴，其中便有这道经典的比亚尼香饭。当锅盖揭开时，第一层白米上点缀着炒至金褐色的洋葱丝与新鲜的薄荷叶，橙黄色的酱汁则是腌制鸡肉时的调料，伴随着扑鼻而来的异域香料气息，混合了红辣椒、番茄、姜黄和蒜蓉的独特风味。用勺子挖下去，看到一层层精心叠加的鸡肉和米饭。萨达尔告诉我，制作传统的比亚尼香饭，肉与饭是分开烹煮的，煮熟后再一层层叠加铺放在一起。这种"层叠法"是制作正宗比亚尼香饭的关键步骤之一。我尝了一口，浓烈的香料在口中层层散开，身体瞬间感到温暖，甚至微微冒汗。萨达尔说，通常巴基斯坦人会用鸡肉或羊肉搭配长粒香米制作比亚尼香饭，但随着人们对健康与可持续发展的关注，越来越多的巴基斯坦人开始选择用蔬菜来代替肉类。

莎夫卡热情地向我传授了一道简化版的蔬菜比亚尼香饭的食谱。所需食材包括：2杯巴斯马蒂长粒香米，各100克的花菜、胡萝卜、豆角和土豆，4个切薄片的洋葱，2个切碎的番茄，1杯酸奶，2汤匙姜蒜蓉，2茶匙红辣椒粉，1茶匙姜黄粉，适量的豆蔻花粉、孜然籽、香菜粉、加兰马萨拉、盐，少许油、薄荷叶、香菜叶，2片香叶和1个柠檬的汁。

准备工作开始时，首先用冷水冲洗长粒香米，直到水变得清澈为止，然后将米浸泡半小时。接着，在平底锅中加热油，将一半的洋葱炒至金黄色，捞出备用。在同样的油中加入剩下的洋葱，炒至半透明，然后加入姜蒜蓉爆香。接着放入切碎的番茄，煮至变软后，再加入蔬菜块和红辣椒粉、姜黄粉、香菜粉、孜然籽、加兰马萨拉以及盐一起翻炒至蔬菜变软，最后加入酸奶搅拌均匀，再煮几分钟。同时，在另一个锅里将水煮沸，加入盐，倒入浸泡好的米，煮至七分熟后沥干水分。将米饭铺在蔬菜上，撒上炸好的洋葱、薄荷叶、香菜叶和柠檬汁。盖上锅盖，用小火焖煮20～30分钟，待米饭与蔬菜的香味完全融合，最后依照"层叠法"将饭菜交替铺放。

我对莎夫卡说："蔬菜版的比亚尼香饭都需要花费不少时间和心力呢。"她微笑着回答："是的，因为这是你对家人、朋友表达爱意的方式呀！"

巴基斯坦比亚尼香饭
Pakistani Biryani Rice

Pakistani Vegetable Biryani

Saying that Biryani has become the culinary icon of Pakistan isn't an exaggeration, it's a delicious fact. Actually, the spices and ingredients used in this classic dish are the main elements that make up Pakistani food landscape. This dish, packed with more than ten local spices, is a feast for the senses. It's like getting a VIP pass to a flavor festival. Furthermore, it tells the story of South Asia's rich culinary tradition and culture.

The origin of Biryani can be traced back to the Mughal dynasty in South Asia (the Mughal dynasty ruled most of South Asia from 1526 to 1858). The word "Biryani" comes from the Persian word "Birian", which means "deep fried before cooking". This technique was used by royal chefs in Mughal times to infuse maximum aroma into each grain of rice. Some historians believe that the invention of the Biryani is attributed to Mumtaz Mahal, the wife of Mughal Emperor Shah Jahan, who built the Taj Mahal. One time, she noticed that there were many malnourished soldiers in the army, ordered a dish made with nutritious and appetizing spices, meat, and rice, and thus Biryani was born. From then on, the status of Biryani in the entire South Asian subcontinent, including Pakistan, India, Sri Lanka and other countries and regions, has been quickly on the rise to its close to legendary status today. In present-day Pakistan, Biryani isn't just a meal, but the main character at special occasions such as weddings, Eid al-Fitr celebrations, and family gatherings.

Once I was invited to dinner at the home of the Pakistani Consul General in Guangzhou Sardar Muhammad. That day, the Consul General's wife, Shaftqa, dressed in Pakistani traditional costume, laid out a feast for us, featuring her exquisite mutton Biryani as the highlight. When the lid was lifted, the top layer of white rice was dotted with golden-brown fried onions and green mint leaves, while patches of brown orange marinated mutton peeked through. The air was thick with the mixed aroma of spices, chili peppers, tomato, turmeric, and garlic. Spooning into it, I discovered the Biryani's signature layers: rice, mutton, rice and more meat. Consul General Sardar explained that traditional Biryani involves cooking the meat and rice separately, then layering them together for the final dish. As I took my first bite, a whirlwind of spices swirled in my mouth, warming me from the inside out. Sardar mentioned that typically, Pakistanis use long-grain basmati rice with chicken or mutton for Biryani. But recently, there's been a shift towards vegetable Biryani, thanks to a growing focus on health and sustainability.

I asked Shaftqa for a simplified vegetarian Biryani recipe, and here's what she shared. For ingredients you'll need 2 cups of long-grain basmati rice, 100 grams each of cauliflower, carrots, green beans, and potatoes, all diced; 4 onions, thinly sliced; 2 tomatoes, chopped; 1 cup of plain yogurt; 2 tablespoons of ginger-garlic paste; 2 teaspoons of red chili powder; 1 teaspoon of turmeric powder; cardamom powder; cumin seeds; 2 teaspoons of coriander powder; 1 teaspoon of garam masala; salt; oil; a handful of mint leaves (pudina); a few sprigs of cilantro; 2 bay leaves; and the juice of one lemon.

Once you've gathered all the ingredients, start by rinsing the long-grain basmati rice in cold water until it runs clear. Soak the rice in water for about half an hour. In a large saucepan, heat some oil and fry half the sliced onions until they're golden brown. Remove them from the oil and set them aside. In the same oil, add the remaining onions and cook until they're translucent. Stir in the ginger-garlic paste and saute until fragrant. Next, add the chopped tomatoes and cook until they soften. Toss in the mixed vegetable cubes along with the red chili powder, turmeric powder, coriander powder, cumin seeds, garam masala, and salt. Cook over medium heat until the vegetables are tender. Stir in the yogurt and let it cook for a few more minutes. Meanwhile, in another pot, bring some water to a boil and add a pinch of salt. Drain the soaked rice and add it to the boiling water. Cook the rice until it's about three-quarters done, then drain any excess water. Layer the partially cooked rice over the vegetables in the saucepan.

Sprinkle the top with the fried onions, mint leaves, cilantro, and a squeeze of lemon juice. Cover the pot and cook on low heat for about 20-30 minutes, until the rice is fully cooked and the flavors have melded together. Use the layering technique to combine the rice and vegetables for serving.

I told Shaftqa that even making a vegetable Biryani takes a good deal of time and effort. She smiled and said, "Yes, but that's how you show your love for family and friends, isn't it?"

做客巴基斯坦驻穗总领事萨达尔·穆罕默德的家，品尝比亚尼香饭
照片来源：《Lingling探世界》节目
Tasting Biryani at the residence of Mr. Sardar Muhammad, Consul General of Pakistan in Guangzhou
Photo Source: *See the World with Lingling* TV Program

我的父母在家中厨房做饭
My parents cooking together in our home kitchen

第 4 章
Chapter 4

这些佳肴烹饪起来节能、省时又省心
Gourmet Dishes Which are Energy-Efficient,
Time-Saving, and Worry-Free

　　在这一章里面，我将着重给大家介绍在烹饪过程中节能或者省事、省时的美食菜肴以及其背后的历史典故、人文趣事。粤菜美食文化博大精深，傲古骄今，单论粤菜饮食文化中的烹饪技法，就有将近70种之多，而这众多烹饪技艺之中，有不少操作起来简单、节能、低碳又美味的技法。在这一章节里我会给大家介绍清蒸、刺身处理、炖、白灼、炒、炸等快手烹饪方式。

　　广州是一座极其包容的城市，当地人对美食从不挑剔。无论是本土出品的佳肴，还是来自其他省市乃至世界各地的美食，他们都心怀好奇，乐于尝试。因此，在这里，总能品尝到各种各样的美味。自有民谚"食在广州，美食天堂"可不是随便说说的，颇有海纳百川之蓬勃气势。所以，本章除了介绍烹饪过程低碳、节能的粤菜以外，也包含了数道国外美食。用广东的食材、国外的做法，让融合中外的菜肴在广州这座享有盛誉的国际美食大都市里尽展所长。

In this chapter, I will focus on introducing food dishes that save energy and time in the cooking process, as well as the stories and history behind them. Cantonese food culture is extensive and profound, making locals proud in the past and present. There are nearly 70 kinds of cooking techniques in Cantonese food culture, many of which are simple, energy-saving, and result in delicious dishes. In this chapter, I will specifically touch on quick cooking methods such as steaming, stir-frying, roasting, and others.

Guangzhou is an extremely inclusive city where locals are never picky about food. Whether it is the local dishes or foods from other provinces and even around the world, they are always curious and eager to try. Thus, one can always savor a diverse array of flavors here. There is a local saying which roughly translated means, "Eat in Guangzhou, A Food paradise." To give credit to the international foods that have made the Guangzhou food scene diverse, this chapter not only introduces the low-carbon and energy-saving cuisine in Cantonese cuisine but also includes several foreign dishes.

10秒搞定姜撞奶，驰名粤港澳

　　粤菜，作为世界美食文化中历史悠久、底蕴深厚的菜系之一，以其菜式繁多、技法精湛而闻名。有的时候，粤菜厨师们为了成就一款美味而不吝啬烦琐工序，只为出品能够打动人心。不过，在粤菜中，也有一些制作过程极其简单，耗时短暂，却同样令人欲罢不能的经典小吃。姜撞奶，就是这么一道甜品，其最早源自广东番禺。它香醇爽滑，甜中微辣，风味独特。而其制作难度低、原料简单，使得它成为能在家轻松制作的首选小吃。姜撞奶的主要原料仅为小黄姜、牛奶和冰糖，绝对是馋了、饿了时不求人的好选择。因此，我将它作为此章介绍的第一道美食。

　　我曾在黄埔古港古村品尝到地道的粤式风味，其中就包括这道姜撞奶。千年黄埔古港，昔日的繁华虽已沉淀于历史长河，但近年来并未沉寂，反而焕发出新的生机，愈发热闹非凡。除了古色古香的历史遗迹，最吸引游客的便是那里的美食飘香。古村入口两旁的小吃店鳞次栉比，陈列着琳琅满目的传统小吃：鸡仔饼，老婆饼，老公饼，牛杂，甜酸萝卜，凉拌鱼皮，蛋挞，双皮奶，姜撞奶……令人目不暇接。我有幸请教当地的"奶婆"，她耐心分享了在家制作姜撞奶的秘诀。

　　姜撞奶最大的一个特点就是做好以后呈凝固状态，放个勺子上去也不会塌。而要成功制作出一碗完美凝固的姜撞奶，关键在于了解其凝固原理。牛奶中含有一种叫作酪蛋白的蛋白质，而生姜汁中的生姜蛋白酶能够分解酪蛋白，使得牛奶凝固。而生姜蛋白酶在70℃左右的温度凝乳效果最佳，所以我们在制作姜撞奶时需要先将牛奶加热至70℃。选用黄姜作为食材，有助于凝固。制作时，先将黄姜榨出约1小勺姜汁，倒入碗中。随后将加热到70℃的牛奶高高举起，缓缓冲入姜汁中，静置5分钟左右，待其自然凝固。表面光滑，放上勺子不塌，则说明姜撞奶已制作成功，入口即化，香甜嫩滑。

姜撞奶特别适合在春季食用。在湿气重的回南天，多数时候人感觉昏昏沉沉的，食欲不振。然而，姜撞奶里面的姜辣素能刺激胃肠黏膜，有利于食物的消化和吸收，提神醒脑，尤其对女性而言，是一道难得的养生美味。

广式姜撞奶制作中
Making Cantonese Ginger Milk Pudding

广式姜撞奶
Cantonese Ginger Milk Pudding

Ginger Milk Pudding—Just Ten Seconds to Make the Greater Bay Area's Famous Dessert

Cantonese cuisine, on the world stage of cuisine, has one of the longest historical and cultural heritages, showcasing a wide variety of dishes and colorful cooking techniques. Cantonese chefs often go to great lengths, using complicated procedures to create delicious dishes that move people's hearts. However, there are also famous dishes in Cantonese cuisine that take a short time to cook and are addictive and unforgettable. One such snack is ginger milk pudding, originally from Panyu, Guangdong. It is fragrant, refreshing, smooth, sweet, and slightly spicy, with a unique flavor. Among all Cantonese dishes and snacks, this famous dish is the easiest to make, requiring only three raw ingredients: ginger, milk, and rock sugar. It can be made at home, even when time is short, without any assistance. Therefore, I introduce it as the first delicacy in this chapter.

Previously, I visited the ancient village of Huangpu and sampled many local Cantonese flavors, including the ginger milk pudding. The old Huangpu Port has thrived for over a thousand years, becoming increasingly vibrant and showcasing its centuries-old charm to visitors. Besides the famous scenic spots, the most alluring aspect is the fragrant cuisine. At the entrance of Gugang Village, an array of snack shops offers various delicacies: chicken biscuits, wife's cake, husband's cake, beef offal, sweet and sour radish, cold fish skin, egg tarts, double skin milk, ginger milk pudding, and more. I sought tips from the local "milk lady" (also known as Nai Po) on how to prepare ginger milk pudding at home.

The key characteristic of ginger milk pudding is its ability to solidify without breaking, even when a spoon is placed on it. To successfully prepare this pudding, it's important to understand the factors that lead to its solidification. Milk contains a protein called casein, while ginger juice contains ginger protease, which breaks down the surface protein of casein particles, causing the milk to solidify. The best coagulation effect of ginger protease occurs at a temperature of around 70°C(158°F), so when making ginger milk pudding, the milk should be heated to 70°C(158°F). Yellow ginger is recommended, which aids the solidification process. To prepare, squeeze a small spoonful of ginger juice from the turmeric and place it in a bowl. Then, pour the hot milk into the bowl from a height, and after the collision, let the liquid stay in the bowl for about 5 minutes to solidify. If the skin is not broken when a spoon is placed on it, the pudding is successfully done. This authentic ginger milk pudding offers a silky sensation in one's mouth that melts instantly upon ingestion.

Ginger milk pudding is a great choice during the damp spring weather in South China. Many people feel lethargic and lose their appetite during this time, but the gingerol in ginger milk pudding can stimulate the gastrointestinal mucosa, aiding in the digestion and absorption of food, and providing a refreshing effect. This dessert is also particularly beneficial for women's health and beauty.

七彩鱼生助你捞到风生水起

食用生鱼片的传统可追溯至中国古代，早在先秦时期便已有记载。2 000多年前，孔子曾提出"食不厌精，脍不厌细"的观点。这里的"脍"指的是将新鲜的鱼类或贝类切片后蘸调料生吃的食物，"不厌细"强调了鱼片要切得越薄越好。由此可见，在孔子的时代，人们已对鱼生极为喜爱且颇有研究。后来，食用鱼生的习惯在隋唐时期传至日本、韩国，逐渐演变为日本国菜之一的"刺身"（Sashimi），这已经是尽人皆知的事实。

在中国广东的顺德一带，乡民们一向偏爱以脆肉鲩鱼片搭配各类配菜制作七彩鱼生。后来，当地人逐渐对七彩鱼生菜谱做出改良，除了用脆肉鲩鱼，还会使用三文鱼。据说，这种变化是因为随着人们对健康的关注度日益增加，人们开始担心脆肉鲩是淡水鱼，而淡水常被视为"死水"，即便使用杀虫药水处理，仍无法完全确保鱼肉中没有寄生虫，如肝吸虫等。这种顾虑使得生活在童话世界般纯净海域中的挪威三文鱼更加受到青睐。因为海水是"活水"，有天然的杀虫作用，能够更好地保障鱼肉不被污染。因此，越来越多的广东餐厅开始使用三文鱼制作七彩鱼生。

后来，这道充满顺德粤菜特色的鱼生传入新加坡、马来西亚等地，以赤、橙、黄、绿的艳丽色彩成为当地华人春节期间的热门菜肴，尤其流行在农历新年的第七天食用。这一天为何如此特殊？这与中国的传统节日习俗息息相关。据传，女娲创世时在第七天创造了人类，因此正月初七被定为"人日"，即是人类的生日。生日庆祝自然要配以喜庆的菜肴，七彩鱼生因其亮丽的色调和吉祥的寓意而成为首选菜。此外，身处异国他乡的华人过年时往往会聚在一起，共同品尝家乡的美味，说一些吉利话。七彩鱼生在享用前需要用筷子搅拌均匀才更美味，大家各自拿起一双公筷，一边搅拌一边祝愿："一捞运气，二捞财气，三捞福气，捞起捞起，捞到风生水起！"在粤语中，"捞得起"意指赚大钱、发大财。当地华人见这句话吉利，于是又将此菜肴改名为"捞起鱼生"。

　　自己在家做捞起鱼生亦十分简单。只要将鱼生和一众配菜材料准备好，加上花生油就可以食用。实则就是刺身，无须烹饪。具体步骤为：在超市购买薄切的三文鱼生片，配上酸姜段、葱丝、香菜、红萝卜丝、紫萝卜丝、青瓜丝、酸藠头、花生碎、紫姜椒，充分搅拌后放入花生油，有些人还喜欢调制一小碟沙拉酱混合芥末油来蘸食。一道色香味俱全的佳肴就这样呈现于你面前——适合在与亲友欢聚时享用，寓意大家在新的一年里"捞到风生水起"，赚到盆满钵满。

"意头菜" 捞起七彩鱼生
Lao Qi Seven Colors Raw Fish (Pick Up Salmon Raw Fish)

Salmon Raw Fish with Assorted Cantonese Snacks and Pickles—A Lucky and Festive Dish

The practice of consuming raw fish has its origins in ancient China, dating back to the Pre-Qin period. Over two thousand years ago, Confucius advocated for savoring the essence of food and enjoying thinly sliced fish. This preference for thinly sliced fish is evident in the tradition of eating sashimi. It is clear that our ancestors, as far back as the time of Confucius, held an appreciation for raw fish and established standards for its preparation. Subsequently, the Chinese custom of consuming raw fish made its way to Japan, the Republic of Korea, and other regions during the Sui and Tang dynasties.

In Shunde, Guangdong Province, villagers have always enjoyed preparing raw fish with grass carp, combining it with various side dishes and ingredients to create the dish known as "Seven Colors Raw Fish". Over time, the locals enhanced the recipe by incorporating salmon in addition to the grass carp. This change was influenced by the perception that salmon is a "healthier" option, as it is believed that freshwater fish like grass carp may contain parasites due to the stagnant water in freshwater aquaculture. In contrast, Norwegian salmon, living in the pristine seawater, is considered relatively cleaner and less likely to be contaminated. As a result, many restaurants in Guangdong now use salmon instead of grass carp to prepare "Seven Colors Raw Fish".

The dish "Seven Colors Raw Fish", a popular Shunde Cantonese dish, was later introduced to Singapore, Malaysia, and other places. It became a favored Chinese New Year dish due to its vibrant red, orange, yellow, and green colors, symbolizing prosperity and vitality. This dish is traditionally consumed on the seventh day of the Chinese New Year. The significance of this date is twofold. Firstly, according to legend, the seventh day of the lunar calendar is considered the birthday of mankind, as it is believed that Nü Wa created humans on this day. To celebrate this occasion, festive dishes are prepared, and the colorful "Seven Colors Raw Fish" with its auspicious colors is a fitting choice. Secondly, during the Chinese New Year, despite being in a foreign land, Chinese communities gather to enjoy their traditional cuisine and exchange auspicious greetings. The ritual of "lao"ing the "Seven Colors Raw Fish", where everyone stirs the dish with their chopsticks, is accompanied by the recitation of phrases symbolizing luck, wealth, and blessings. The term "lao" also connotes the accumulation of wealth and good fortune in Cantonese. This auspicious tradition led to the renaming of the dish to "Lao Qi Raw Fish" (Pick Up Raw Fish) by local overseas Chinese communities.

The dish is simple to make at home. Thinly sliced salmon is paired with sour ginger, shredded scallions, coriander, shredded red radish, shredded purple radish, shredded cucumber, pickled leek, crushed peanuts, and purple ginger pepper. After mixing well and adding peanut oil, some people also prepare a small plate of white salad dressing with mustard oil for dipping. This colorful and flavorful dish is traditionally enjoyed with family and friends during gatherings, symbolizing good fortune for the days to come.

清蒸海上鲜

　　蒸，源自我们的先祖，是中国独有的烹饪法之一，其他国家甚少使用这一技法，有也多半出现在国外的唐人街。蒸的过程十分简单，只需要将食物架在盛水的器皿上，待水烧至沸腾，借助水蒸气的热量，将食材蒸熟。蒸的妙处在于，食物不与器皿直接接触，但能够在烹饪过程中吸收适量的水分，使得成品既嫩滑鲜美，又保持了食材本身的原汁原味。

　　在广东，能吃到各式各样的海鲜，且它们大多适合清蒸。常见的蒸鱼包括海鲈鱼、淡水鲈鱼、黄花鱼、小黄鱼和鲩鱼等。其中，清蒸海鲈鱼尤其受欢迎，并已成为广东的传统名菜之一。海鲈鱼分为白鲈鱼和黑鲈鱼两种，富含蛋白质、维生素A、B族维生素及钙、镁、锌、硒等营养元素。清蒸海鲈鱼只需要用葱段和姜花去腥，烹饪时间几分钟，耗能低；鱼蒸出来以后香甜鲜美、细嫩爽滑，深受老广们的喜爱。

　　这道菜的具体做法是：将处理干净的海鲈鱼置于盘中，底部垫上葱条、香菜，上面铺上姜片，然后放入已经烧沸水的竹蒸笼中蒸熟。

　　细心的读者可能会问，为什么要用竹笼蒸而不是随意找一个铁锅来蒸呢？这其中大有讲究。金属材质的锅盖因其导热快、散热也快，容易在锅内外形成温差，当水蒸气遇冷后迅速凝结成水滴滴回食物上（俗称"水倒滴"），不仅稀释了食材的鲜味，还破坏了菜品的外观。而使用竹编盖则不会有这样的温差问题，能够有效避免水滴落入食物，保持食物的色香味俱佳。

　　我们还可以通过"蒸"这个字来解读其中奥妙。中国的汉字是象形文字，"蒸"

字带了个草花头而非铁字旁，这也恰巧佐证了我们的先祖在发明"蒸"的时候用的是植物编织而成的盖子而非别的；使用竹篾编织的盖子不容易形成内热与外冷的温差，可以避免食物在被热蒸汽致熟的同时出现水倒滴的情况。只有使用正确的蒸法，才能呈现出最地道的清蒸海鲈鱼风味。

广东主妇们在拿捏蒸鱼烹饪时间上非常精准，鱼蒸久了鱼肉容易过硬，时间不够则外熟内生、吃起来不够卫生。如果想测试鱼是否已经蒸熟，可以在揭盖后用筷子轻轻插入鱼身，若能轻松穿透，便表示鱼已经蒸得恰到好处。

之前采访过一位资深粤菜师傅，他说，传统的粤式清蒸一般还会在碟子里再铺上冬菇丝和火腿丝，但是这种清蒸法多用于我们平日里吃到的塘鲜、河鲜。河鲜、塘鲜或多或少会带有"泥味"，放了冬菇丝和火腿丝能去除这些味道并帮助提鲜。但海鲈鱼是海鲜，本身就带独特鲜味，所以清蒸鲈鱼时大多只用葱段和姜花，待到鲈鱼熟后浇上蒸鱼豉油即可。

清蒸海鲈鱼
Steamed Sea Bass

Steaming Cantonese Fish

The cooking method of "steaming" has its origins in ancient China and is one of the distinctive cooking techniques of Chinese cuisine. This method is not commonly found in other countries and is often prevalent in Chinatowns abroad. Steaming is a simple process that involves placing a food rack over boiling water in a pot or wok, allowing the steam to cook the food. The beauty of this method lies in its ability to gently infuse moisture into the food without direct contact with the cooking vessel. As a result, the cooked food is tender, smooth, and retains its freshness.

In Guangdong, you can enjoy a variety of seafood, including sea bass, pond-raised bass, yellow croaker, small yellow croaker, and grass carp, which are commonly used for steaming. A very popular dish is steamed sea bass, a traditional favorite in Guangdong. Sea bass is divided into two types: white bass and black bass, both rich in protein, vitamin A, B-complex vitamins, calcium, magnesium, zinc, selenium, and other nutrients. Steamed sea bass only requires scallions and ginger flowers, and the cooking time is just a few minutes, resulting in low energy consumption. Once the fish is steamed, it has a sweet and delicious taste, and is tender and smooth, making it deeply loved by Guangdong people.

The specific method is to place the cleaned sea bass on a plate, arrange scallions, coriander, and ginger flowers on top, and then steam it in a bamboo cage with boiling water beneath. This method is commonly used for steaming fish and is a popular way to prepare sea bass in Cantonese cuisine.

Careful readers can't help but ask, why put a sea bass in a bamboo cage instead of randomly finding a regular metal pot to steam it? This is because metal lids absorb heat quickly but also dissipate heat quickly, making it easy to form a temperature difference between the internal environment and external environment. When steam meets such metal lids, it quickly condenses into water droplets and drips onto the food (a phenomenon known as "water dripping"), blurring the freshness of the food and affecting its aesthetics. Bamboo steamers, on the other hand, prevent such temperature imbalances and avoid unwanted water from dripping onto the food, preserving the dish's texture and taste.

We can also decipher the right way of making steamed fish through how the Chinese character "steaming" (蒸) is written. Chinese characters are logographics, the word "steaming" comes with a "grass flower" top (in the character) instead of an "metal" side, suggesting that our ancestors used a lid woven from plants, such as bamboo, to craft their steamers. The use of bamboo woven covers does not easily create a temperature difference between internal and external environments, avoiding the occurrence of water dripping while food is cooked by hot steam. And this is also the correct way to steam a sea bass.

Guangdong housewives are skilled at preparing steamed fish. If the fish is steamed for too long, the meat tends to become tough. If the steaming time is insufficient, the fish may be undercooked, posing a risk to health. To check if the fish is fully cooked, you can use chopsticks to pierce the fish after removing the lid. If the chopsticks can easily penetrate the fish, it is cooked and ready to serve.

Previously, I interviewed a senior Cantonese chef who said that the traditional Cantonese style of "steaming" usually involves laying shredded mushrooms and ham on the plate underneath the fish. However, this steaming method is often used for the freshwater fish we eat. The freshwater products from ponds and creeks may have a "muddy taste" to some extent, and adding shredded mushrooms and ham can eliminate these flavors and help improve freshness. But sea bass is seafood with a unique flavor, so when steaming sea bass, most people only use scallions and ginger. After the sea bass is cooked, it can be topped with steamed fish soy sauce.

广东人的腊味情怀

俗话说："秋风起，食腊味。" 每当秋意渐浓时，广州的大街小巷便弥漫着腊味的香气。腊味店门前人头攒动，排起长队，许多广州人趁此机会采购腊味，或赠亲友，或为自家食用。不少家庭也会自行腌制腊肉，走过居民楼时，常常能见到屋顶悬挂着整排的腊肉，色泽诱人，散发出独特的风味。腊味是一种非常适合长期储存的食品。平时只需用纸包好，放入冰箱，便可保存两三个月。

许多人家煲饭时，爱在米饭中放几块腊味一同蒸煮，让肥美的油香渗透进米粒，也流入心田。这不仅是饱腹之食，更是一种温暖的情感联结。然而，鲜为人知的是，采用清蒸的方式，更能保留腊味的原始风味，使其更加醇香。将洗净、切片的腊肉或腊肠，搭配芋头或粉葛，放入蒸锅中，用清水蒸约20分钟，待熟透即可。如果想减少盐分摄入，则无须再加盐或酱油，腊味的自然咸香会在蒸煮过程中渗入芋头，蒸好后香气四溢，令人垂涎。

根据历史记载，广式腊肠的起源可以追溯到唐朝。当时，中东商人来到广东经商，带来了他们的灌肠工艺。聪慧的广东厨师从中获取灵感，逐步研究出适合本地口味的腊肠制作方法。最初，腊肠是普通家庭为避免浪费而自制的。人们将吃剩的肉剁碎，加以粤式调味再填入猪肠皮中，慢慢便形成了如今的广式腊肠。广式腊肠不仅继承了中华传统的腊制技艺，还融合了外来的灌肠工艺，堪称中外饮食文化碰撞的结晶。

值得一提的是，还可以尝试使用酒水浸泡保存腊肠的独特方法。将腊肠切块，放入酒中浸泡10~20分钟，然后密封于玻璃罐中，存放1周后，腊肠便可作为零食、小吃或佐酒之物，随时享用。这样浸泡过的腊肠不仅味道更好，而且带有酒的醇香。

广州人家院子里晒的腊肉以及腊肠
Cured meats and sausages drying in a Guangzhou courtyard

The Cantonese Affection for Cured Meat

As the saying goes, "When the autumn wind blows, you eat preserved meat." Every year in autumn, the streets and alleys in Guangzhou are filled with the aroma of cured meat, and lines are long in front of cured meat shops. Many people in Guangzhou take this opportunity to stock up on these delicacies, either as gifts for family and friends or for their own consumption. Many people even make their own cured meats at home. Walking through neighborhoods, one can often spot rows of colorful cured meats hanging on the roof of the building appealing with tempting colors. Cured meat is an easy-to-store food. Wrapped in paper and placed in the refrigerator, it can last 2-3 months.

Many families enjoy steaming rice together with a few pieces of cured meat for homemade meals. The moist and rich flavor infuses into the rice, warming the hearts of those who partake and maintaining an emotional bond with family and friends. Little do they know that steaming cured meats on their own can preserve their original flavor more effectively than steaming them together with rice. After washing and slicing the cured meat, pair it with taro or kudzu for about 20 minutes. The natural saltiness of the cured meat will infuse the taro, creating a dish that's both fragrant and flavorful. If you're watching your sodium intake, there's no need to add extra salt or soy sauce—the cured meat itself will provide all the flavor.

According to historical records, Cantonese cured sausages originated in the Tang Dynasty. Merchants from the Middle East brought their sausage-making techniques to Guangdong. The resourceful Guangdong chefs were quick to adapt these methods, evolving them into what would eventually become the Cantonese-style cured sausages we know today. Initially, these sausages were made by common households to avoid wasting leftover meat. Guangdong people chopped up leftover meat, seasoned with local flavors, stuffed it into pig intestines, and gradually forming today's Cantonese-style cured sausages. Over time, it became a unique fusion of traditional Chinese sausage curing techniques and foreign influences—a testament to the culinary exchange between China and the wider world.

It's worth mentioning that another intriguing method is to preserve sausage in alcohol. Slice the sausage into chunks and soak them in wine for 10-20 minutes, then store them in a sealed glass jar for a week. After that, the sausage is ready to be enjoyed as a snack, appetizer, or side dish for drinks. The alcohol not only enhances the flavor but adds a subtle, aromatic complexity to the sausage, making it taste even better.

小镇男孩的卡卢卢鱼

很早以前，卡卢卢鱼就是安哥拉沿海城镇里家家户户都爱做的一道美食。安哥拉位于非洲西南部，西临大西洋，有着丰富的鱼虾类海产。海边小镇的家庭因容易捕捞鱼、虾、蟹等海产品，所以即便经济并不富裕，家家户户也都不缺海鲜。小镇的奶奶、妈妈们将从海里捕捞回来的鱼晒干以后，切成块，再配上番茄酱汁、洋葱、秋葵以及绿叶蔬菜等烹饪成一道特色菜——卡卢卢鱼。

我第一次吃卡卢卢鱼，是受安哥拉前驻穗总领事若昂·达科斯塔先生以及夫人的邀请，到他们府邸做客并共进晚餐时品尝到这道菜肴的正宗风味的。传统的卡卢卢鱼用海鱼干和一些必需配菜——秋葵、绿叶子蔬菜、棕榈油以及番茄酱汁来制作。那天，我们的晚餐由总领事先生从非洲带过来的厨师，一位名叫皮肯奴的非洲帅小伙儿掌勺，他对这道菜进行了自己的诠释与创新，鱼肉一半用的是从安哥拉带过来的海鱼干，另一半用的是在广州当地采购的新鲜马鲛鱼。我问他使用新鲜鱼肉有何意义，他说，新鲜鱼块肉质嫩滑，鱼干的咸鲜味则比较突出，对于吃惯清淡口味的老广们来说可能偏咸，然而鱼干与新鲜鱼块一起烹饪后不仅会在一定程度上稀释海鱼干的浓稠咸味，而且让新鲜的马鲛鱼块更加入味，使菜肴口感达到平衡。他还在菜肴里加入了中国大葱。

我有幸进到总领事家的后厨观赏这道菜肴的制作过程，烹饪步骤比想象中容易。安哥拉卡卢卢鱼所用食材包括150克切成块的海鱼干、150克切成块的新鲜海鱼块、150克菠菜、50克秋葵、2个番茄和1个洋葱，调料包括有棕榈油、大蒜、中国大葱、浓缩番茄酱以及少量的鱼味调味料。皮肯奴先把番茄、洋葱切块，大蒜和中国大葱切碎，与浓缩番茄酱一起放入榨汁机里搅拌成泥；接着，他用棕榈油起油锅，放入之前做好的酱泥，再放入切好的菠菜段、秋葵，加入几块鱼味调味料之后倒入所有鱼块，用文火煮上20分钟即大功告成。

卡卢卢鱼的异域香味芬芳四溢，香味飘满了整个饭厅，与白米饭搭配食用特别下饭。据总领事夫人珍妮介绍，以前，卡卢卢鱼只在安哥拉沿海地区的城镇流行，也是近二三十来年才在安哥拉的内陆城市广泛普及。总领事更提到，以前安哥拉国内的交通并不方便，加上经历了27年之久的内战，许多铁路、公路遭到损坏。在内战结束以后，有幸得到中国政府的大力帮助而兴建了本格拉铁路，促进了城乡间人民的往来，让许多海边小镇的青年更容易去到大城市打工，同时把不少家乡菜引入内陆城市。

我在想，这道卡卢卢鱼或许就是小镇男孩引入大城市的菜肴之一。背井离乡的他，在闲暇之余思念亲人，于是按照家里祖母、妈妈传给他的秘方去制作这道卡卢卢鱼。吃上一口，仿佛家人近在咫尺，犹如回到了海边小镇，闻到了那令人无比熟悉的咸海水味道，缓解了乡愁。有意思的是，皮肯奴就是一名小镇青年，来自安哥拉的海边小镇洛比托，早年去到安哥拉的首都罗安达打工，后来几经辗转又来到国际化大都市广州。我总觉得，他就是那许许多多把卡卢卢鱼引入大城市的小镇男孩之一。只不过，借着中国与安哥拉的友好往来，他让生活在广州的我也有机会品尝到了这道安哥拉佳肴。

安哥拉卡卢卢鱼
Angolan Calulu de Peixe Fish

A Small Town Boy's Calulu Fish

Calulu de Peixe Fish has long been a beloved dish in the coastal towns of Angola. Because Angola's western coast being adjacent to the Atlantic Ocean, it was easy to catch fish, shrimp, and crabs. Despite the modest economic conditions in the villages, seafood was abundant. The grandparents and mothers in the seaside towns of Angola would catch sea fish, dry them, cut them into pieces, and cook them with delicious tomato sauce, onions, okra and green leafy vegetables to make Calulu Fish.

My first encounter with the rich flavors of Calulu fish was at the gracious dinner invitation of Ambassador Joao Da Costa, the former Consul General of Angola in Guangzhou, and his wife Jeanne. Traditionally, Calulu is cooked using dried saltwater fish and included key elements such as okra, leafy greens, palm oil, and a robust tomato sauce. That night, the Consul General's chef Pikeno cooked Calulu for us and he infused the traditional recipe with his own creativity, using half of the saltwater fish he brought from Angola, dried grouper, and the other half was fresh mackerel purchased from a local Guangzhou market. I asked Pikeno about his new recipe of mixing fresh fish with dried fish. He explained that the fresh fish offered a tender, succulent texture, while the dried fish contributed a more pronounced, salty depth. For those Guangzhou locals accustomed to a lighter and more subtle flavors typical of Guangzhou cuisine, the intense saltiness of dried fish alone might be too much. However, when simmered together, the fresh mackerel absorbed just enough of the dried fish's essence, achieving a harmonious balance of flavors. Additionally, Pikeno added Chinese onion to the dish, which was not part of the original Calulu fish recipe.

I observed the cooking process in the Consul General's kitchen, and was pleasantly surprised it was easier than expected. The ingredients for Angolan Calulu fish include 150 grams of dried fish, cut into pieces; 150 grams of fresh fish, also cut into pieces; 150 grams of spinach; 50 grams of okra; 2 tomatoes; and 1 onion. The seasonings comprised of palm oil, garlic, Chinese spring onion, concentrated tomato puree and a few cubes of Maggi Seasoning Fish Stock. Pikeno began by chopping the tomatoes, onion, garlic, and Chinese spring onion, blending them into a smooth paste with concentrated tomato puree. He then heated palm oil in a pan, added the tomato-onion paste, followed by the spinach, okra and a few fish seasoning cubes. Finally, he simmered the mixture with all the fish for 20 minutes, allowing the flavors to meld beautifully.

The exotic aroma of Calulu filled the dining room, with its flavors paired perfectly with steamed white rice. Jeanne, the Consul General's wife, shared some insights into the history of the dish. She explained that Calulu was once a cherished specialty mostly in the coastal towns of Angola, only becoming popular in the inland cities over the past two or three decades. Ambassador Dacosta added that Angola's domestic transportation was once quite limited, exacerbated by the damage caused by the country's 27-year civil war. However, after the civil war ended, with the help of the Chinese government, Angola built the Benguela Railway, promoting exchanges between urban and rural areas. Many young people from coastal towns went to work in big cities such as the capital Luanda, bringing their hometown dishes with them.

As I savored each bite, I imagined the journey of this dish from a small seaside town to a bustling city—a young man, far from home, finding solace in recreating his grandmother's recipe. Each bite eased their homesickness, making it feel as if their families were close by and they could smell the familiar saltiness of their hometown in the air. Interestingly, Pikeno hails from a coastal town in Angola named Lobito. He moved to the capital, Luanda, and eventually made his way to the international metropolis of Guangzhou. I can't help but think of him as one of the many village boys who have brought the authentic Calulu fish recipe to the big cities in Angola. Thanks to the friendly exchanges between China and Angola, he has given us the opportunity to taste this extraordinary dish right here in Guangzhou.

地道的冬阴功汤烹饪起来竟然如此快捷和简单

泰国驻穗总领事馆提供的数据显示，目前，共有10 000多名泰国籍人士在广州长期生活以及工作。他们中的大部分主要在进出口贸易、餐饮以及教育领域谋事。

通过一位朋友的介绍，我认识了在广州从事餐饮行业的撒拉贡。撒拉贡是一位泰国人，说话斯文且待人彬彬有礼。他在广州开了一家泰式餐馆，隐匿在黄金地段的小街小巷之中，但却时常座无虚席。据说，撒拉贡在东南亚有很多赚钱的生意，他开这家店是出于情怀，主要目标不在于赚多少钱，而是在广东做出地道正宗的泰国菜。

有一次，我问撒拉贡："广州本地人最喜欢的泰国菜是什么？"他说，绝对要数冬阴功汤了。

有一次我去撒拉贡的家中做客，目睹他下厨烹饪这道冬阴功汤的过程。只见他在一锅事先准备好的虾汤里快速地放入香茅、南姜、柠檬叶等香草料，用大汤勺在锅里不断搅动、翻腾大约3分钟，然后加入草菇和小红辣椒，待汤汁再次煮滚后放入虾肉和鱼露，随即熄火，淋上青柠汁调味，用香菜叶做装饰。整个过程，用时不超过5分钟。用时之少、动作之快，令人咋舌。

新鲜出炉的冬阴功汤香气扑鼻，此外，红的辣椒、绿的香茅，各种颜色相互映衬，令人赏心悦目。第一口喝下去，感觉汤里所包含的各种味道同时在我口中一点点地迸发出来，酸的酸、辣的辣、甜的甜，层次分明，各种口感相互交融，让人从中品尝到了生机勃勃的大自然所蕴含的希望。

我多次去泰国的经验告诉我，一道传统的泰国菜肴里通常会包含有甜、酸、辣、咸、甘苦等几种泰式基本口味。如果食客在吃下去第一口的瞬间就能够辨别出里面所混合的各种味道，那就是极品。如果吃进嘴里只是一味地让人感觉到辣，或是酸味太重，那就未免有些落俗了。

关于冬阴功汤的来历还有一个小故事。相传，泰王郑信的女儿淼运公主曾经得了一种疾病，对任何东西都食之无味。后来，御厨特别调制了这款刺激味蕾、有助开胃的冬阴功汤，公主喝着喝着，胃口好转，病也痊愈。自那时起，冬阴功汤便有了泰国国汤的美誉。时至今日，在泰国，冬阴功汤依然是上至王公大臣，下至黎民百姓都常食用的一道汤肴。

泰式冬阴功汤
Thai Tom Yum Goong Soup

Authentic Tom Yum Goong: Surprisingly Quick and Easy to Make

According to data provided by the Consulate General of Thailand in Guangzhou, there are currently over 10 000 Thai nationals who have been living and working in Guangzhou. Most of them work in the fields of import and export trade, catering, and education.

Through a friend's introduction, I met Salagon, who is engaged in the restaurant business in Guangzhou. Salagon is a Thai man who is polite and friendly. He opened a Thai restaurant in Guangzhou, hidden in the small alleyways of a prime location, but often full of Thai food lovers. It is said that Salagon has many profitable businesses in Southeast Asia, and this restaurant is just his passion. His main goal is not to earn much money, but to produce authentic Thai cuisine in Guangdong.

Once I asked Salagon,"your restaurant has served many locals in Guangzhou, what is their favorite Thai dish?" He said that it definitely has to be this bowl of Tom Yum Goong soup that keeps bringing people back.

Later, I visited Salagon's home and witnessed him cook this Tom Yum Goong with my own eyes. He added some sliced galangal, chopped lemongrass, and coriander root into a boiling shrimp stock. He stirred the pot with a large spoon for about 3 minutes, then added straw mushrooms, kaffir lime leaves and small red chili peppers. Once the stock was simmering, he added deveined prawns and fish sauce. He then turned off the heat and added lime juice and cilantro leaves. The entire process took no more than 5 minutes. The efficient use of time and speed of his movements were breathtaking.

The freshly made Tom Yum Goong is fragrant, with the red chili and green galangal, creating a visually pleasing array of colors. With the first sip, the various flavors in the soup burst out in the mouth, with sour, spicy, and sweet tastes obvious but blending, giving a feeling of hope and vitality.

From my numerous trips to Thailand, I've learned that a traditional Thai dish typically embodies a harmonious blend of sweet, sour, spicy, salty, and even bitter flavors—each a fundamental taste of Thai cuisine. The true mark of a culinary masterpiece should allow diners to identify the various basic flavors in the first bite or sip. If a dish only vaguely hints at spiciness or sourness, then it's not the real deal.

There's also an interesting tale about the origin of Tom Yum Goong. It is said that Princess Miao Yun, the daughter of Emperor Zheng Xin, fell ill, lost her appetite, gradually becoming emaciated. Given Thailand's tropical climate, with its year-round heat and humidity, this was even more cause for concern. In an effort to help her, the Emperor's chef created the Tom Yum Goong, which was both nutritious and tantalizing to the taste buds. Miraculously, after eating the soup, the princess regained her appetite and made a full recovery. As a result, Emperor Zheng Xin declared Tom Yum Goong the national soup of Thailand. To the present day in Thailand, it remains a beloved dish enjoyed by people from all walks of life, from royalty to the common people.

打抛猪肉碎，泰国街头的"金不换"美味！

在写这本书的时候，有朋友问我："除了蔬果、豆类和海鲜之外，你还会推荐大家吃肉吗？"毕竟，肉类，尤其是红肉的碳排放量比蔬菜和豆制品都高，从环境保护的角度来看确实稍次一档。但是我想，任何事情都有一个度的把握，不是说碳排放量高的肉类就完全将其拒之门外，而是应该有选择地吃，少吃一点。这样一来既可做到满足口腹之欲，也能做到营养摄取更加均衡。所以在这个小节，我给大家推荐一款做起来耗时极少的猪肉泰国菜。

众所周知，泰国菜是世界上最受欢迎的菜式之一，且口味也被老广们所接受。我在泰国品尝过不少地道美食，其中泰国的街头美食绝对让人印象深刻。走在人行道上，不时见到成排的路边摊，场面颇为壮观。小贩们从家中带来所需炊具，摆摊揽客，有的吹起口哨、哼起小曲，抑或摇铃吆喝一番，乐此不疲。各样美食，琳琅满目，充分刺激着行人的食欲。而食物的香味夹杂着泰国特有的辣味，混合着各种东南亚香草的味道，充满了异国情调。

有机会去泰国旅行逛街"叹"美食，有一款碟头饭配菜是一定要试的——打抛猪肉碎，它做起来容易，口感丰富且已经成为泰国的街头名菜。打抛猪肉碎的主要食材包括猪肉碎、翠绿色的"金不换"、大小红辣椒以及各种酱料。"金不换"其实就是菜肴名称里的"打抛"，它属于罗勒叶类香草的一种，有着一种异乎寻常的香味，甜中带有浓郁的辛香味，有点类似九层塔，但与九层塔相比，辛香味更重一点。"金不换"除了给菜肴增添甘甜香味使其更下饭外，还具有祛风利湿、散瘀消炎等作用。

　　后来，在一次受泰国驻穗总领事孔雀丽女士邀请前往她府邸品尝美食的时候，我再次品尝到了这款打抛猪肉碎。那天，看着她从泰国带过来的厨师亲自下厨为我们烹饪美食。厨房的桌子上是早已经准备好的各种材料：大蒜、大红辣椒、小红辣椒、红葱头碎、蚝油、鱼露、泰式淡酱油、泰式甜酱油、热炒油、猪肉碎200克以及"金不换"。只见厨师起油锅，放入大蒜、大小红辣椒以及红葱头碎不断翻炒，待香味出来时放入猪肉碎。将猪肉碎炒到七八成熟以后，加入泰式淡酱油、蚝油、鱼露以及一点点的泰式甜酱油，急翻快炒，最后放入"金不换"。

　　看着猪肉碎吸收饱酱油后颜色迅速由浅变深，褐色中点缀着辣椒的红与"金不换"的绿，特别显眼。没几分钟时间，一碟浸足了功夫的打抛猪肉碎就已经做好上桌。打抛猪肉碎在泰国经常以碟头饭的形式来呈现，盛碗白饭在碟中，外加一个荷包蛋就是经济实惠又营养丰富的一餐。泰国总领事说她每周都会让厨师做上两次给她吃，在泰国，它已经成为从皇亲国戚、社会精英到打工一族、普通老百姓都经常食用的菜肴。

　　目睹总领事的厨师为我们烹饪这道菜的全过程后，我有了在家自己动手做的冲动。毕竟，这道菜所用到的食材经济实惠且在广州获取容易，菜肴烹饪时间短，还用的是我们最熟悉的"炒"。在家做了几次以后，慢慢地吃上瘾了。后来每次去泰国，经过正在烹制打抛猪肉碎的摊档，我都忍不住停下脚步，站在那里张望，即便不吃，看看、闻闻也是好的。

泰式打抛猪肉碎
Thai Stir-Fried Minced Pork with Basil Dish (Pad Kra Pao)

The Irresistible Street Food of Thailand—Stir-Fried Pork with Holy Basil

While writing this book, a friend asked me, "Besides vegetables, fruits, legumes, and seafood, would you recommend eating meat?" After all, meat, especially red meat, has a higher carbon footprint compared to vegetables and plant-based foods, making it less ideal from an environmental perspective. However, I believe that moderation is key. It's not about completely cutting out high-carbon foods, but rather making informed choices and consuming them in moderation. This way, we can still enjoy the flavors we love while maintaining a balanced diet. So, in this article, I'd like to introduce you to a simple and quick Thai pork dish that fits the bill.

Thai cuisine is renowned worldwide and is particularly popular in regions like Guangdong, where the bold flavors have found a receptive audience. During my time in Thailand, I indulged in many authentic dishes, but it was the street food that left the most lasting impression. Travelled to Thailand many times and the street food there has left an indelible impression on me. Walking along the sidewalks, I was greeted by rows of bustling street stalls. Vendors brought their cooking equipment from home, set up shop, and welcomed customers with whistles, songs, and even the occasional bell ringing. The variety of food on offer was overwhelming, with aromas that were a tantalizing blend of Thai spices and Southeast Asian herbs, enticing passersby to stop and sample the fare.

If you ever have the chance to visit Thailand, I highly recommend trying a street food dish called Stir-Fried Pork with Holy Basil (known locally as Pad Kra Pao). This dish is simple to make, packed with flavor, and has become a classic on the streets of Thailand. The main ingredients include minced pork, vibrant green holy basil (known as Kra Pao in Thai), red chili peppers, and various sauces. Holy basil, the star of the dish, is a type of basil with a unique flavor that's both sweet and spicy, more intense than Thai basil. In addition to enhancing the dish's flavor, holy basil is also known for its medicinal properties, such as aiding digestion and relieving bloating.

I recall an invitation to the residence of the Thai Consul General in Guangzhou, Madame Krongkanit Rakcharoen, where I had the pleasure of tasting this dish again. On that day, her Thai chef prepared it for us, with the kitchen table laid out with all the necessary ingredients: garlic, large red chili peppers, small red chili peppers, shallots, oyster sauce, fish sauce, Thai light soy sauce, Thai sweet soy sauce, cooking oil, 200 grams of minced pork, and holy basil leaves. The chef began by heating oil in the pan, then adding the garlic, large and small chili peppers, and shallots, stirring constantly until the aroma filled the kitchen. Next, the minced pork was added, cooked until almost done, and then seasoned with Thai light soy sauce, oyster sauce, fish sauce, and a touch of Thai sweet soy sauce. The final touch was adding the holy basil.

As I watched the minced pork absorb the sauces, turning from light to a rich brown, with the red chilies and green basil creating a visually stunning contrast, my mouth watered in anticipation. In just a few minutes, a fragrant plate of Stir-Fried Pork with Holy Basil was ready to be served. In Thailand, this dish is typically presented as a "one-plate meal" with a serving of rice and a fried egg, making for an affordable, nutritious, and satisfying meal. The Consul General mentioned that she has her chef prepare this dish for her twice a week, as it has become a beloved dish for everyone, from the Thai royal family and social elites to the working class and ordinary citizens.

Seeing the Consul General's chef skillfully prepare this dish inspired me to try it at home. The ingredients are inexpensive and easily found in Guangzhou, the cooking time is short, and the method—stir-frying—is one I'm very familiar with. After making it a few times, I found myself getting hooked on it. Now, every time I return to Thailand and pass by street stalls cooking Pad Kra Pao, I can't help but stop, even if just to catch a whiff of the familiar aroma or to watch the process.

快手烹饪甜酸咕噜虾

甜酸咕噜虾是一道健康又简单的粤菜，它衍变自顺德名菜甜酸咕噜肉，除了把猪肉换成九节虾以外，甜酸酱料和配菜基本不变。甜酸咕噜虾采用山楂汁和番茄酱调配成酱汁，吃一口到嘴里，甜甜酸酸、醒神开胃，极受欢迎。

说到甜酸咕噜肉，就不得不提它的起源。相传清朝咸丰年间，顺德陈村因其得天独厚的经商地理位置，吸引了大批洋人到此。洋人们爱上了当地一道名叫"甜酸生炒骨"的菜肴，喜其酸酸甜甜的口感，但他们吃不习惯带骨头的肉，所以，陈村的厨子索性将骨头去掉。洋人吃了以后觉得味道不错，于是竖起大拇指夸赞说了句："Good!"（"赞"的英文单词，与"咕噜"同音。）不懂英文的陈村厨子们听完还以为是在说"咕噜"，因而给这道菜起名为"咕噜肉"。

19世纪中期，广东移民将甜酸咕噜肉引入美国，并且因为山楂取材困难，所以改用当地特色食材番茄酱，这让甜酸咕噜肉的品质得到改良。在吸收了美式元素以后，这道菜谱又传回国内。如今，几乎每家粤菜馆烹饪的甜酸咕噜肉都会用到番茄酱。

前面说到，甜酸咕噜虾和甜酸咕噜肉的制作材料除了所用的肉不同以外，其他基本一样。虾在养殖过程中的碳排放量比猪肉要低，而且营养丰富，天生带有鲜味，仅需配料两三样，且烹饪耗时低，所以使用虾肉而非猪肉还能为减低碳排放量贡献一份力量。

这道菜肴的所需材料包括鲜九节虾、番茄酱、米醋、鸡蛋黄、菠萝罐头、青红椒、洋葱、砂糖、浓缩山楂汁、生粉、盐。首先，准备半斤新鲜九节虾，以还在游水的最佳。挑选时，用网去捞游得快的虾，其越快则表明越强健、越新鲜；捞起来后，看一下色泽和虾壳——有光泽的虾，肉质比较鲜美，虾壳硬且与肌肉之间连接紧密、不易用手剥开的较为新鲜。用剪刀剪去虾头和虾尾，搁置一旁待用（下个章节会具体介绍如何利用"虾废料"）。将虾壳完全分剥干净后，用片刀在虾背上划一刀，不要将肉斩断，方便稍后炒成虾球。将虾放在厨房纸或者一块干净毛巾上面沥干水分。接着将虾放回碗里，加些许盐、生粉以及半个鸡蛋黄搅拌调味。起油锅，放入虾煎炸至金黄后捞出。另起一个干净的锅，烧热后加入米醋、砂糖、番茄酱、浓缩山楂汁，再加少许生粉勾芡。把虾球放回锅炒均匀，再加入罐头菠萝粒、青红椒和洋葱片翻炒一下即可。

甜酸咕噜虾酸甜适中、味道诱人，且不似咕噜肉那般油腻，特别适合小孩、老人食用，开胃下饭。

我曾有机会拜访粤菜厨师曾文洪先生。曾文洪是米其林一星级厨师，我向他请教甜酸咕噜虾的酱汁调配。他说，酱汁是决定一道菜肴成功的关键，可以尝试加入OK汁（即噎汁，粤菜中常用的微辣甜酸汁）和山楂片去调配，这样的酱汁口感更丰富。

炸虾
Fried Prawns

甜酸咕噜虾
Sweet and Sour Gulu Prawns

A Fast Cooking of Sweet and Sour Gulu Prawns

Sweet and Sour Gulu Prawns is a healthy and simple Cantonese dish that evolved from Shunde's famous Sweet and Sour Gulu Pork. Apart from replacing pork with tiger prawns, the other ingredients, including sweet and sour sauce and bell peppers, remain the same. Sweet and Sour Gulu Prawns is made with hawthorn juice and ketchup. Take a bite, it's sweet, sour, refreshing, and appetizing, which makes it a true crowd-pleaser.

Speaking of Sweet and Sour Gulu Pork, one cannot overlook its origins. It's said that during the Xianfeng era of the Qing Dynasty, the Village of Chen in Shunde became a hub for foreign merchants due to its strategic location. The foreigners fell in love with a local dish called "Sweet and Sour Fried Pork Bone", but were unaccustomed to eating meat in a dish which had not been de-boned. So the chefs in the Chen Village removed the bones. After

tasting the boneless version, the foreigners gave it a thumbs-up, exclaiming, "Good!"—a word that sounds like "gulu"(咕噜)in Chinese. The chefs, unfamiliar with English, thought the foreigners were saying "gulu", hence led to the dish being called "Gulu" Pork in Chinese by the local chefs. The name stuck and is still in use to this day.

In the mid-19th century, Cantonese immigrants introduced Sweet and Sour Gulu Pork to the United States, but due to the difficulty in finding hawthorn berries, they switched to a local specialty ingredient—ketchup—to provide the tangy sweetness in the dish. After absorbing these American elements, the recipe returned to China. Nowadays, almost every Cantonese restaurant uses ketchup as one ingredient to make their Sweet and Sour Gulu Pork.

As mentioned earlier, the recipe for making

Sweet and Sour Gulu Pork and Gulu Prawns is basically the same, except for the protein used. The harvesting of prawns emits less carbon than raising pork, prawns are nutrient-rich, they take less time to cook, and are generally regarded as healthier than pork, which makes them perhaps a wiser choice than pork for this dish.

The ingredients you'll need for this dish include fresh tiger prawns, ketchup, rice vinegar, egg yolk, canned pineapple, green and red bell peppers, onion, sugar, concentrated hawthorn juice, cornstarch and salt. Start by selecting half a kilo of fresh tiger prawns—purchased alive if available in your area, for ultimate freshness. When selecting fresh prawns, use a net to catch the fastest and stronger ones. Inspect the color and prawn shell; shells that are hard and not easily peeled off by hand are considered the most fresh. Cut off the heads and tails of the prawn with scissors and set them aside (the next chapter will explain how to make the best use of "prawn waste"). After the prawn shell has been removed, use a knife to make a long slice along the back of the prawn. Do not cut completely through the meat as this will help the prawn turn it into a ball shape during frying. Place

the prawn on kitchen paper or a clean towel to drain. Then put the prawn back into a bowl and add some salt, cornstarch, and half of the egg yolk; stir and season. Heat up a pan or oil, fry the prawn until golden brown, and then remove the prawn from the pan and put aside. In a new clean pan, add the rice vinegar, sugar, ketchup, concentrated hawthorn juice, and a little cornstarch to thicken. Next add the prawn balls to the pan and stir well. Finally, add the canned pineapple, green and red bell peppers and onion slices, stir fry well.

Sweet and Sour Gulu Prawns is characteristically sweet and sour, enticing, and is less greasy than Gulu Pork. As such, it's especially suitable for children and the elderly, and also makes a great appetizer.

Previously, I had the opportunity to visit Cantonese chef Mr. Zeng Wenhong, a Michelin-Star chef. I asked him for tips on perfecting the sauce for Sweet and Sour Gulu Prawns. He emphasized that the sauce is key to the dish's success and suggested trying a blend of OK mixed sauce and hawthorn slices to nailing the authentic sweet and sour flavor profile.

省心炖一碗咸柠檬芫荽瘦肉汤

在广东，自来就有喝靓汤的习俗，每家每户几乎每一天都有煮汤的习惯。从小我最喜欢喝的就是妈妈做的炖汤，她还和我说，古人自有"一滚、二煲、三炖"的说法，即炖汤要达到最佳效果需要3小时。虽然耗时较长，但"炖"是将所有食材放在密封的容器内煮，能最大程度地存留食物本身的香味，炖出来的汤味道尤其鲜美；加上汤被维持在80～85℃，营养破坏较少。现在，在家做炖汤，只需要一个家庭电子炖盅和两三样食材即可，炖的时候亦无须人看守，不必担心水被烧干，省心省力。

在广东，素来都有用山泉水炖汤的习惯。之前有机会走访粤菜专门店山泉公馆，老板吴浩坚喜欢琢磨用山泉水与不同原材料进行搭配去炖各种滋补汤水。他觉得现代人生活节奏快，最好能多喝一些润燥祛火的汤饮，还特别独创了咸柠檬芫荽瘦肉汤。在粤港澳大湾区以及东南亚部分地区，向来流行用咸柠檬搭配七喜做成咸柠七特饮。虽然咸柠七生津润肺，但是毕竟加入了碳酸饮料，加上多是冻饮，所以不够健康。吴老板和我说，咸柠檬芫荽瘦肉汤营养且健康，很受老饕们欢迎，尤其是生活节奏快的年轻人。

我看它做起来容易，所以特别向吴老板讨教烹饪配方。此汤所用到的食材不多，包括腌制2～3年的咸柠檬2小块、芫荽头连须1条、精瘦肉100克、鸡爪1个、桂圆肉3粒以及白胡椒1粒。炖汤讲究材料新鲜、纯净以及保持恒温。先将所有材料洗净放入炖盅里面，然后用保鲜膜紧紧裹住炖盅以防香味溢出，再将炖盅放进蒸锅里，通过蒸汽加热。炖汤讲究的是科学的烹饪理念，不用超过100℃的高温或者滚烫的水去制作，以避免食物里面的营养被破坏，蒸汽的最高温度就是100℃，可将汤料维持恒温在90℃左右——所有营养和好东西都是慢慢地被炖煮出来的。

广东当地人经常食用咸柠檬。后来发现，用盐腌制柠檬可以对柠檬进行储存、优化、增鲜以及提香。除此之外，咸柠檬还可以去火、祛寒。所以在顺德一带，经常能看到有不少民居里放着腌制10年以上的咸柠檬，这已成为传统，当地人也以此为豪。

咸柠檬芫荽瘦肉汤材料
Ingredients of Cantonese Pork Soup with Salted Lime and Coriander

咸柠檬芫荽瘦肉汤
Cantonese Pork Soup with Salted Lime and Coriander

Effortless Delight—Cantonese Pork Soup with Salted Lime and Coriander

In Guangdong, there has been a long-standing tradition of enjoying flavorful soups, with almost every household having the habit of cooking soup nearly every day. From a young age, my favorite was always my mother's slow-cooked soups. She would often remind me of the old saying"one boil, two simmers, three slow-cooked", meaning that to achieve the best flavor, the soup should be slow-cooked for three hours. Though a little time-consuming, this cooking method, where ingredients are sealed in a pot surrounded by simmering water of around 90°C (194°F), allows the natural aromas and flavors to be preserved, making the soup particularly delicious. Additionally, because the soup is kept at a gentle cooking temperature of around 80-85°C (176-185°F), the nutrients remain largely intact. Nowadays, making a slow-cooked soup at home is much

easier with an electric slow cooker. With a few ingredients, during the slow cooking process of 2-3 hours, no monitoring is required and there is no risk of the water evaporating, making it a time and energy-efficient cooking method. This long-standing Cantonese tradition produces delectable and fragrant slow-cooked soups with minimal effort.

In Guangdong, there has always been a tradition of slow-cooking soups with mountain spring water. I previously had the opportunity to visit Shanquan Mansion, a Cantonese restaurant which specializes in this technique. The owner, Mr. Wu Haojian, enjoys experimenting with using mountain spring water combined with different ingredients to slow-cook various nourishing soup broths. He believes that modern people lead fast-paced lives, so it is best to drink more hydrating and "fire dispelling" (cooling) soups. He created an innovative salted lemon and cilantro slow-cooked lean pork soup. In the Guangdong-Hongkong-Macao Greater Bay Area, as well as some parts of Southeast Asia, a popular "Salted Lime Seven Up" drink is made by mixing salted lime with 7-Up soda. Although the salted lime seven up drink moistens the lungs and throat, it also contains carbonated soda, which is not particularly healthy, especially when served over ice as a cold beverage. Mr. Wu explained to me that his salted limes and coriander pork soup is a healthier option and flavorful, and quickly became extremely popular among Cantonese gourmets, especially young people with fast-paced lifestyles.

This is a flavorful and nutritious soup that can be enjoyed year-round, and is known for nourishing and hydrating dry throats and lungs. I asked Mr. Wu for the recipe. The simple ingredients needed are 2 small pieces of salted lime, preserved for 2-3 years, 1 coriander plant with roots attached, 100 grams lean pork, 1 chicken foot, 3 dried longans, and 1 white peppercorn.

The emphasis is on using fresh, pure ingredients and maintaining a constant low temperature while slow-cooking. First, wash all ingredients and place them in a ceramic pot or bowl. Next, tightly cover the pot with plastic wrap to seal in the aromas. Then place the covered pot in a steamer to cook with steam heat. The key is using the scientific slow-cooking method, avoiding high temperatures over 100°C (212°F) or boiling, which can damage nutrients. Since the maximum temperature of steam 100°C, it's a perfect choice. After steaming, the soup is kept at a constant low temperature around 90°C (194°F). This allows the flavors and nutrients to gently release into the broth over 2-3 hours of slow simmering.

Guangdong natives often eat salted limes. Using salt to preserve limes can provide storage, optimization, freshness, and aroma enhancement. As such, in the Shunde region, it is common for many households to have salted limes that have been marinating for over a decade. Possessing these aged, salted limes has become a local tradition that residents take pride in.

简单易做的中东小吃——鹰嘴豆泥酱

　　鹰嘴豆泥酱（胡姆斯）是一道来自中东地区的特色美食，主要食材包括鹰嘴豆、芝麻酱、蒜和柠檬，做法并不复杂，但是味道甘美、营养健康，所以深受不同族裔人群的喜爱。

　　到底是谁发明了胡姆斯，今天已无从考究。虽然希腊人、土耳其人、以色列人、叙利亚人等多个地方的人都声称他们是胡姆斯的发明者，但却没有足够的证据。鹰嘴豆这种豆类食材的起源在中东地区可以追溯到10 000多年前，而鹰嘴豆泥酱（胡姆斯）最早出现在13世纪开罗编写的食谱中。在13世纪的阿拉伯烹饪书*Kitab Wasf al-Atima al-Mutada*中出现了被称为"hummus kasa"的芝麻酱：它以鹰嘴豆和芝麻酱为基础原料，并且包含了许多香料、草药、坚果和大蒜等。

　　多年前，在受邀前往以色列前驻穗总领事劳霈乐先生家做客的时候，我有机会品尝到了这道颇具传统风味的美食。以色列位于中东地区，同时也是三大洲——欧洲、亚洲与非洲的交界处。劳霈乐告诉我："在以色列，走进任何一家餐厅或者当地人的家里，都能见到这道具有特色的胡姆斯。"（劳霈乐从事外交行业已经30余年，曾经在广州出任以色列驻穗总领事长达4年的时间。）以色列主要人口是犹太人，其余的是阿拉伯人以及少数其他族裔。虽说不同的信仰往往导致饮食上的巨大差异，但是胡姆斯却是犹太人、阿拉伯人以及其他族裔家庭餐桌上都经常出现的美味。劳霈乐告诉我，他在广州工作期间想念家乡菜时，做得最多的就是胡姆斯。胡姆斯是他的爸爸教会他做的，他的爸爸很会做菜，在做菜之余也鼓励家中的孩子们多参与烹饪。小时候，他的爸爸常常带着他和兄弟姐妹一起制作胡姆斯：每一位家庭成员拿起一样不同的食材，逐一将它们放进搅拌机里——妹妹放孜然粉、哥哥放西班牙甜胡椒粉……直到放完所有食材才进行搅拌。制作胡姆斯成为一家人共享天伦之乐的活动之一，劳动的同时增进了家人间的感情。

那天，总领事邀请我与他一起下厨制作这道胡姆斯。我从他手中接过来浸泡开了的鹰嘴豆放进搅拌机里；然后再放入各种辅料——黏稠的中东白芝麻酱，用烘焙过的白芝麻制作而成，闻起来芬芳扑鼻；色泽鲜艳的西班牙烟熏甜味胡椒粉，让人想起弗拉门戈女舞者身上那一抹艳丽的鲜红披肩；中东混合香料粉，提神醒脑胜过一杯咖啡。随后，总领事又将孜然粉、百里香粉、蒜片等逐一放进搅拌机，一阵搅拌机颤动的声响之后，胡姆斯做好了。

劳霈乐为我们准备了中东皮塔（pita）卷饼以及中国馒头。因为，鹰嘴豆泥的最佳吃法是与各种食物搭配着来吃，包括面包、米饭、面条、沙拉，乃至蔬菜、胡萝卜条、西芹条、玉米片等。我拿起勺子挖了一大勺涂抹在pita卷饼上，咬了一口，绵绵的，入口即化，胡姆斯里的各种调料散发着浓郁的清香，咀嚼过后口里回甘。

我问劳霈乐，为何他做的这道酱比我在其他任何一个地方吃到的都要好吃。他先是谦虚地回答说，这是我一起参与了制作这道酱菜的缘故。在我的再三追问之下他才说，在胡姆斯上浇淋足够多的橄榄油是关键。的确，我记得他在搅拌好的胡姆斯上浇淋了很多很多的橄榄油，直到漫过酱泥3厘米左右。劳霈乐说，加入大量橄榄油后的鹰嘴豆泥酱在冰箱内存放一两周都没有问题，但是，由于胡姆斯太好吃了，所以通常做一次，不到两天就吃完了！

胡姆斯做起来耗时少，所用食材不涉及任何肉类，健康美味且容易消化，乃是一道不可多得的绿色、低碳营养素食。

鹰嘴豆泥酱
Hummus

Simple and Delicious Middle Eastern Snack—Hummus

Hummus is a quintessential delicacy from the Middle East region, flavorful and nutritious made with chickpea as the main ingredient, together with tahini, garlic, and lemon. Simple to prepare, its rich taste and health benefits have made it beloved by people of all different religions and backgrounds in the region.

The origins of hummus are shrouded in mystery, with no definitive answer as to who first created this delectable dish, with multiple claimants— Greeks, Turkish people, Israelis or Syrians— vie for the honor of its invention, but none offer conclusive proof. Chickpeas, the main ingredient, have been a staple in the Middle East region for over 10 000 years, and the origin of hummus can be traced back to the 13th century in a recipe from Cairo. In the 13th century Arabian cookbook *Kitab Wasf al-Atima al-Mutada*, a sesame paste called "hummus kasa" is recorded: it is made with chickpeas and tahini, infused with

a variety of spices, herbs, nuts, and garlic.

Years ago, I was invited to the residence of Mr. Peleg Lewi, the former Consul General of Israel in Guangzhou, where I had the chance to savor the traditional flavors of this dish. Israel, situated at the crossroads of Europe, Asia, and Africa, has a rich culinary heritage deeply connected with the region. Peleg shared with me that in Israel, hummus is a staple found in almost every restaurant and household. (A diplomat with over 30 years of experience, Peleg served as Israel's Consul General in Guangzhou for 4 years.) Despite the varied dietary practices due to different religious beliefs in Israel, hummus remains a beloved dish shared amongst Jews, Arabs, Christians and other ethnic peoples alike. Peleg fondly recalled how, during his time in Guangzhou, he often made hummus whenever he missed the flavors of home. It was his father who taught him the art of making hummus, who was a skilled cook and encouraged all the children to participate. At home, his father would often gather him and his siblings to make hummus together:each family member would pick up a different ingredient and add it one by one to the blender—his sister the cumin powder, his brother the Spanish sweet paprika, until all the ingredients were in the machine. Making hummus became one of the bonding activities for family members enjoy together.

That day, in Peleg's home, he invited me to make hummus with him. I took the soaked chickpeas from him first and put them in the blender. Then I added a variety of other spices and herbs—the thick tahini paste made from baked white sesame seeds smelled so fragrant. There was also the bright, colorful paprika, reminding me of a flamenco dancer's vibrant red shawl. The Middle Eastern mixed spice powder gave it a nice kick, better than a cup of coffee. After me, Peleg put in the rest of the ingredients, and after the whirring sound of the blender, we had the hummus!

For us guests, Peleg thoughtfully provided pita rolls and Chinese steamed bread buns to accompany the hummus. The ideal way to savor hummus is by pairing it with like bread, rice, noodles, salads, or even crunchy vegetables such as carrot sticks, celery sticks, and corn chips. I eagerly grabbed a spoon, generously spread a large dollop of the hummus onto a warm pita bread, and took a bite. The creamy texture melted blissfully in my mouth, instantly releasing an aromatic medley of spices that enveloped my senses with a profound sense of happiness.

I asked Peleg why this hummus he made was better than any hummus I'd ever eaten anywhere else. He replied humbly that it was because I helped make it. Then he said that pouring enough olive oil onto the hummus is the key. Indeed, I remember that day he poured a lot of olive oil over it until it was about 3 centimeters deep. More olive oil also helps keep hummus fresh. The chickpea puree with a lot of olive oil can be stored in the fridge for a week or two and still be fresh. But Peleg then said, with a tongue-in-cheek smile, that it wouldn't last that long usually because the hummus is so delicious, he would eat it all day long, and it would be all gone before he knew it.

Hummus is a quick and easy dish to make, with no meat, and is a healthy, tasty, and easy on the stomach when comes to digestion—a rare green, low-carbon and nutritious vegetarian treat!

干杯
Cheers.

做客以色列前驻穗总领事劳霈乐的府邸
照片来源：《Lingling探世界》节目
Dining at the residence of Peleg Lewi, former Israeli Consul General in Guangzhou
Photo Source: *See the World with Lingling* TV Program

莫吉托 —— 治愈身心的鸡尾酒

一首由周杰伦演唱的单曲《莫吉托》（英文名为"Mojito"），唱响了大江南北。莫吉托是一种鸡尾酒的名称，主要材料包括朗姆酒、青柠汁、薄荷叶、糖与苏打水。它源自拉丁美洲的古巴，酒精度数较低——酒精含量为10%左右，加之入口清爽，所以在世界各地都颇受年轻人的喜爱，算得上是世间最有名气的鸡尾酒之一。

之前认识了一位居住在深圳的古巴小伙儿汉塞尔，他说，莫吉托做起来十分容易。首先，准备好朗姆酒、鲜榨青柠汁、薄荷叶、糖、冰块以及苏打水，然后用勺子或者（如果有条件的话）用研杵和研钵将青柠汁、薄荷叶和糖一起轻轻碾碎，再与朗姆酒、其他配料一起倒入酒杯里，充分搅匀后就可以喝了。

据说，莫吉托的前身是一个叫"德雷克"的酒，它是世界上第一杯鸡尾酒，由大海盗弗朗西斯·德雷克发明。16世纪，德雷克从英国开启了他的环游地球之旅，到古巴附近时感觉身体不适，于是将船停泊在古巴，上岸寻觅可以治愈他和船员的草药。当地人告诉他们用薄荷叶、青柠片、甘蔗树皮以及朗姆酒可以治疗晕船呕吐，于是他将这些材料全数混在一起，调配出了号称世界上第一杯鸡尾酒的德雷克。将德雷克酒的配方减去甘蔗树皮，加入蔗糖浆、苏打水和冰块，便成了莫吉托。

我问汉塞尔做好一杯莫吉托的关键是什么。他说，关键在于要像对待初恋一样温柔地准备此酒，搅拌的时候充满了爱与柔情。

古巴莫吉托鸡尾酒
Cuban Mojito

Mojito—A Passionate Drink from Cuba

Previously, a song titled "Mojito" performed by the renowned Chinese singer Jay Chou became a nationwide hit. Mojito, is also the name of a cocktail crafted from 5 key ingredients：rum, lime juice, mint leaves, sugar and soda water. Originating from Cuba, this cocktail boasts a low alcohol content of around 10%, and its refreshing taste has made it popular among young people and has risen to become one of the most famous and iconic cocktails around the world.

I once met a young Cuban man named Hansel, who was living in Shenzhen working as a bartender. He shared with me that making a Mojito is really quite simple. Let me share the cocktail recipe：gather rum, freshly squeezed lime juice, mint leaves, sugar, ice cubes, and soda water. Gently muddle the lime juice, mint leaves and sugar using a spoon or, if you have it, a mortar and pestle. Then, pour this mixture

along with the rum and other ingredients into a glass, give it a good stir, and it's ready to be enjoyed.

It is said that the Mojito's predecessor was a cocktail called a "Drake". Considered the world's first cocktail, it was developed by the renowned explorer Sir Francis Drake. Back in the 16th century, while Drake embarked on his circumnavigation of the globe from England. When he and his crew fell ill near Cuba, they anchored on the island to search for remedies. From the locals, they learned that mint leaves, lime slices, sugarcane bark, and rum had therapeutic effects, so they combined all these ingredients, purportedly creating the world's first cocktail, which was appropriately named after Drake himself and gradually became a hugely popular drink among the locals. "The Drake" cocktail formula, with the sugarcane bark replaced by sugarcane syrup, eventually evolved into the modern Mojito.

I asked Hansel what the secret is to making the perfect Mojito. He smiled and said prepare the drink as if you were tending to your beloved one—stirring it gently, with love and all gentleness.

低碳
行动　绿色美食就在你身边
Low-Carbon Living: Green Cuisine at Your Fingertips

在古巴驻穗总领事玛丽娜的府邸畅享古巴美食
照片来源：《Lingling探世界》节目
Enjoying Cuban cuisine at the residence of Marina Mylnikova, Consul General of Cuba in Guangzhou
Photo Source: *See the World with Lingling* TV Program

广东炒饭，越变越好吃

　　说起炒饭，其实它算不上是广东的主流菜式。小时候，市面上的食材供应还不甚丰富，家人、朋友们为了不浪费食物，经常会用前一晚剩下的白米饭和鸡蛋、葱花炒在一起，做成一道香喷喷又饱肚的炒饭。当时，感觉它就是天上佳肴一般的美食，特别解馋。后来，随着经济状况的改善，炒饭吃得越来越少。

　　传统广东炒饭的配料及制作步骤并不复杂。在广东，炒饭的做法沿袭至今并没有太大变化。使用隔夜饭或者把大米放在比平日少一点的水中煮熟，尽量使饭粒干爽不粘。起油锅煎蛋，炒到五成熟以后加入米饭、火腿片或者腊肉丁、灼熟的豌豆和胡萝卜丁，最后放入葱花及盐翻炒一下即可。因为炒饭所用食材经济实惠，烹饪起来简单快捷，所以现在我吃它的机会又多了起来。

　　之前因为工作关系，我采访了几任秘鲁驻穗总领事，惊喜地发现，广东炒饭在秘鲁竟然是一道全民美食。当地人把炒饭称为"arroz chaufa"，这个"chaufa"源自"炒饭"的粤语发音。根据秘鲁共和国驻华大使馆提供的资料，广东炒饭是最早为秘鲁食客所熟知的中餐菜品。19世纪中期，超过10万广东移民远渡重洋到达秘鲁，成为当地劳工。为了节省时间和金钱，他们经常将剩下的米饭、肉和一些蔬菜炒在一起。慢慢地，这道省时、省力又饱肚的菜肴便在当地流传开来。炒饭被广东移民引入秘鲁以后，其配方被改良以更好地迎合当地人的口味。其中，传统广东炒饭的火腿片被换成了鸡腿肉丁、牛肉丁或者是烤鸭肉丁，当地人还会放入海虾仁；另外，在秘鲁吃到的炒饭一律会放入酱油，这和秘鲁人爱吃重口味的菜肴有关，所以米饭看上去是浅褐色的，而非原有的米白色或者米黄色。

我曾经受秘鲁驻穗总领事鲁本·埃斯皮诺先生与他的夫人珍妮邀请，到他们家做客，在他的家中品尝到了他太太亲自下厨制作的秘鲁版炒饭。这款炒饭所用到的食材和配料与秘鲁当地的制作方式基本一样。将鸡蛋搅拌成蛋液后加入盐和胡椒粉，鸡肉丁炒至五成熟后撒少许盐与黑胡椒粉，腊肠丁、新鲜海虾仁用酱油拌匀以后煎至微微金黄备用。另外，准备好隔夜饭、葱花、姜碎、花生油、少量芝麻油、酱油。用花生油起油锅，放入少量姜碎以及白色部分的葱花炒蛋，炒到五成熟后加入隔夜饭，不断搅拌翻炒至将米饭疙瘩彻底碾开，再放入腊肠丁、鸡肉丁，然后加入芝麻油继续翻炒，最后放入酱油炒匀，把炒饭盛到盘子里，再摆上海虾仁。

鲁本和我说，秘鲁人真的很喜欢吃炒饭，怎么吃都不腻。他还告诉我："每一位秘鲁人都会做炒饭。除了在秘鲁的唐人街粤菜馆能够吃到炒饭，在秘鲁人的家庭厨房里，每周都会做炒饭来吃。"

秘鲁特色炒饭
Peruvian Arroz Chaufa (Peruvian-styled Fried Rice)

Cantonese Fried Rice, Getting Tastier at Each Turn

Speaking of fried rice, it is not generally considered a mainstream dish in Guangdong. However, when I was young, the supply of ingredients in the market was quite limited, and in order not to waste food, my family and friends often used leftover cooked rice from the previous night, added eggs and scallions and stir fried everything together, to make this delicious and economical dish. At that time, it felt like a heavenly delicacy, and was especially satisfying. Later, as the economic situation improved in China, fried rice was eaten less and less.

The ingredients and cooking steps of traditional Cantonese fried rice are not complicated and the process has not changed much. Use the leftover white rice from the previous day or boil rice in less water than usual; you want rice grains that are more dry and not sticky. Fry eggs in a wok until they are almost cooked, then add rice, sliced ham, diced Cantonese sausage, cooked

peas, and diced carrots. Finally, add scallions and salt and stir fry until golden brown. Because its ingredients are economical with simple and fast cooking steps, I began making fried rice more often at home these days.

Through my interview with several Peruvian Consuls General in Guangzhou, I pleasantly discovered that this seemingly ordinary dish has become a local delicacy in their home country. Local people call fried rice "arroz chaufa" ("chaufa" is derived from the Cantonese pronunciation of "fried rice"). According to the information provided by the Peruvian Embassy in China, fried rice was the first Chinese dish known to Peruvian locals. In the mid-19th century, over 100 000 Guangdong immigrants traversed the ocean to become local laborers in Peru. In order to save time and money, they often stir fried leftover rice, some meat and vegetables. Soon, this time-saving, labor-saving, and delicious dish became popular. After fried rice was introduced into Peru by Guangdong immigrants, it was tailored to better cater to local tastes. The ham slices of the traditional Cantonese fried rice were replaced with diced chicken, beef or roast duck, and local people also add shrimp. In addition, all fried rice eaten in Peru is mixed with soy sauce, which an example of a Peruvian culinary preference for heavily flavored dishes, so fried rice in Peru looks light brown in color.

I was once invited to dinner at the residence of Ruben Espinoza, the Peruvian Consul General in Guangzhou. His wife Jenny made Peruvian-style fried rice. The ingredients used in the fried rice they made were very similar to those used in Peruvian fried rice in Peru. The ingredients and seasonings of this fried rice include beaten eggs, cubed chicken and sausage, and fresh peeled shrimp marinated in soy sauce. This was mixed with leftover rice, scallions, chopped ginger, peanut oil, a small amount of sesame oil and soy sauce. To prepare, heat a frying pan with peanut oil, and saute the chicken cubes until golden, then sprinkle with some salt and black pepper. Put into a bowl for later. In the same pan, use the same process to cook the shrimp and until golden, and add a little soy sauce. Put into a bowl for later use. Heat the same pan with peanut oil, add a small amount of chopped ginger and white parts of shallot and stir fry the beaten eggs until about halfway done. Add the rice and stir fry constantly until it is mixed with the egg. Then add the diced sausages and chicken cubes to the pan, pour a desired amount of soy sauce and a few drops of sesame oil. Saute again, remove from the heat, sprinkle with the green part of shallots, add the cooked shrimp and serve.

Ruben told me that Peruvians really like to eat fried rice and never tire of it. He also said, "most Peruvian people learn to cook fried rice. In addition to eating fried rice in the Chifa (Chinese restaurants) in Peru as a must, fried rice is prepared almost weekly in the Peruvian family kitchen."

做客秘鲁驻穗总领事鲁本·埃斯皮诺的府邸并品尝秘鲁炒饭
照片来源：《Lingling探世界》节目
Sharing Arroz Chaufa at the residence of Ruben Espinoza, Consul General of Peru in Guangzhou
Photo Source: *See the World with Lingling* TV Program

希腊当地每家餐馆的必备——乡村沙拉

在弥漫着神秘气息的希腊，有一道极其适合夏日享用的美食——希腊沙拉。这道菜虽制作简单，却蕴含着浓郁的地中海风情。轻轻一口，仿佛置身于碧海蓝天之间，鼻尖充满了咸咸的海风味道。黑色橄榄与橄榄油散发出浓郁的香气，吃过后口中留有丝丝清甘。

希腊沙拉的原名是"Horiatiki Salata"，直接翻译为"乡村沙拉"或"农夫沙拉"。为什么会有这样一个名字呢？或许读者已经有所了解，古希腊和中国一样，曾是一个农耕社会。大多数人以种地为生，不仅饲养牛，也驯养山羊来耕作。据说，很久以前，希腊的农民和牧羊人为了节省时间和成本，将黄瓜、番茄、洋葱、黑橄榄以及山羊奶酪（即菲达奶酪）用布包好，带到田间或山上。到了中午，他们就用这些食材配上面包一起享用。由于这些食材简单易得，且都是希腊夏季的应季食材，到了20世纪初，希腊的一些餐馆开始将这些食材切块混合，并将其引入餐桌，命名为"希腊沙拉"。由于希腊沙拉价格适中，且清新可口，营养丰富，所以它迅速在希腊各地流行开来。如今，无论你走进希腊的哪一家餐厅，几乎都能看到希腊沙拉的身影，它的声誉也早已传播到世界各地。

有一次，我应希腊前驻穗总领事马汀·曼达里蒂斯的邀请到他府邸做客，他特地为我们准备了这道地道的希腊沙拉。我有幸在厨房目睹了他精湛的厨艺。正宗的希腊沙拉所需的食材包括黄瓜、番茄、洋葱、黑橄榄、菲达奶酪、牛至、黑胡椒和橄榄油，缺一不可。我注意到马汀在制作沙拉时的几个小窍门。切割食材时，尽量使其大小和形状相近，这样视觉上更为美观，酱汁也更容易渗透，食用起来更加方便。为了便于搅拌，他使用了一个大碗作为盛装容器。在制作过程中，食材的加入顺序也很有讲究：首先将番茄连皮切块；然后削去黄瓜的皮，去掉两端后切块；接着是洋葱——剥去外皮和外面两层，只留里面鲜嫩多汁的部分，将洋葱从中间切开，然后像剥花瓣一样一层层地剥下来放入碗中；撒上一点海盐，加入切块的菲达奶酪，放入黑橄榄，撒上牛至和黑胡椒，最后淋上一层厚厚的橄榄油。随后，不断搅拌让食材入味，霎时间，整个厨房弥漫着橄榄油的香气，伴随着浓郁的花果香。马汀说，橄榄油是希腊美食中使用得最多且最重要的元素，它也是地中海饮食的基石。许多希腊人相信，常食用橄榄油有助于延年益寿。他还说，如果你在广州想做一份正宗的希腊沙拉，所有的原材料在本地都能轻松购得，非常便利。

当然，除了上述的食材之外，还可以根据个人喜好加入一些当季的当地食材，如青椒或红椒。有些地方甚至会用樱桃番茄代替普通番茄。不过，生菜叶却从未在希腊沙拉中出现过。虽然生菜是意大利和法国沙拉的常见蔬菜，但在招待客人时，为了保持希腊沙拉的正宗风味，最好还是不要放入生菜叶为妙。

希腊沙拉
Greek Salad

A Staple in Every Greek Restaurant—The Village Salad

In the enchanting land of Greece, there is a dish that perfectly captures the essence of summer— the Greek salad. Simple to prepare, yet rich with Mediterranean flavor, this salad takes you to the azure seas and sun-kissed shores of Greece with just one gentle bite. It's as if the briny taste of the sea lingers in the air as the robust flavors of black olives and olive oil leave a delicate and refreshing aftertaste on you.

The original name of the Greek salad is "Horiatiki Salata", which translates directly to "village salad" or "farmer's salad" . Why such a name? Well, as you may know, like ancient China, ancient Greece was predominantly an agrarian society. Most people lived off the land, raising not only cattle but also goats for farming. It is said that long ago, Greek farmers and shepherds would pack cucumbers, tomatoes, onions, black olives, and goat cheese (known as feta cheese) in cloth bundles to take

to the fields or mountains. At lunchtime, they would pair these simple ingredients with bread for a hearty meal. As these ingredients were easy to obtain and were peak-season foods in the Greek summer, some restaurants in the early 20th century began cutting them into smaller pieces, mixing them together, and serving them on their tables, giving birth to the Greek salad. Its reasonable price, refreshing taste, and nutritional value quickly made it popular throughout Greece. Today, you can hardly step into a Greek restaurant without encountering this beloved dish, which has also earned global recognition.

On one occasion, I was invited to the residence of Martin Mandalidis, former Consul General of Greece in Guangzhou. He graciously prepared an authentic Greek salad for us, and I had the privilege of watching him work his culinary magic in the kitchen. The essential ingredients for a traditional Greek salad include cucumbers, tomatoes, onions, black olives, feta cheese, oregano, black pepper and olive oil—each one indispensable. I noticed a few of Martin's tips for making the perfect salad.When cutting the ingredients, he carefully ensures that they are similar in size and shape, creating a more visually appealing dish and allowing the dressing to coat the pieces evenly, making each bite more enjoyable. He used a large bowl to mix the salad, which made tossing easier and more effective. As for the preparation, there is a certain order to follow:First, cut the tomatoes into chunks without peeling them; then peel the cucumbers, remove the ends, and cut them into similar-sized pieces. Next comes the onion— peel off the outer skin and the first two layers, keeping only the fresh, juicy interior. Slice the onion in half, then separate the layers as if peeling flower petals, placing them into the bowl. Sprinkle a pinch of sea salt, add the cubed feta cheese, and finally, toss in the black olives. Top with a generous drizzle of olive oil, oregano, and a dash of black pepper. A quick stir to mix the ingredients, and soon the kitchen is filled with the rich, fruity aroma of olive oil, mingling with the freshness of the vegetables. Martin mentioned that in Greek cuisine olive oil is the most important element, and it is also the base of Mediterranean cuisine. In Greece, many believe that it is linked to longevity and that's why Greek people use it a lot. He also noted that if you wish to make an authentic Greek salad in Guangzhou, all the necessary ingredients are readily available.

Of course, you can also customize the salad with local seasonal ingredients like green or red bell peppers. Some variations even replace regular tomatoes with cherry tomatoes. However, one ingredient you will never find in a Greek salad is lettuce. Although lettuce is a common component in Italian and French salads, if you want to maintain the authentic Greek flavor when hosting guests, it's best to leave them out.

低碳® 绿色美食就在你身边
行动 Low-Carbon Living: Green Cuisine at Your Fingertips

做客希腊前驻穗总领事马汀·曼达里蒂斯的家，品尝希腊沙拉
照片来源：《Lingling探世界》节目
Enjoying Greek salad at the residence of Martin Mandalidis, former Consul General of Greece in Guangzhou
Photo Source: *See the World with Lingling* TV Program

来一杯比利时巧克力慕斯，给生活增加滋味

　　提到比利时，人们自然会联想到巧克力。比利时巧克力以其独特的口感和卓越的品质享誉全球，这与其精湛的制作工艺和严格挑选的原材料密不可分。比如，一块巧克力中可可粉的颗粒越细，口感就越细腻丝滑，入口即化。而比利时巧克力的研磨技术能将可可粉的颗粒做到极其微小，足以在同行中傲视群雄。比利时的巧克力制造商还自豪地规定，可可粉的颗粒必须小于20微米，通常能达到18微米的标准，这远远优于美国的30～35微米标准。此外，根据政府规定，比利时巧克力中的可可含量必须高于35%，这一标准也高于欧盟规定的25%，有效降低了糖和乳脂等非可可添加物的比例，使巧克力的口感更加浓郁芳香。为了确保纯正的品质，比利时的巧克力制造者只使用纯可可脂，坚决拒绝使用植物油。

　　我曾有幸应邀到比利时驻穗总领事裴伟岷的府邸做客，那天我们讨论了关于可持续发展美食的话题。我请他推荐几道简单易做的比利时美食，他首先想到的就是比利时巧克力慕斯。"太棒了！"我兴奋地说。毕竟，巧克力、华夫饼和啤酒，这些都是比利时享誉世界的代表。

　　据说，在这个仅有1100多万居民的巧克力小王国里，比利时人对巧克力的热爱近乎狂热，他们几乎每天都要吃巧克力，人均年消费量达10.7千克，居欧洲第二，仅次于瑞士的11.6千克。在这个"巧克力王国"里，有超过350家巧克力制造商、2 000多家巧克力店，提供着3 000多种琳琅满目的巧克力产品，可谓竞争激烈，也因此孕育出了不少世界顶级巧克力品牌。

　　总领事告诉我，只要有进口自比利时的巧克力，简单5步就能做出一杯令人垂涎的巧克力慕斯。我兴奋地跟随他走进厨房，打开他家的冰箱，映入眼帘的是满满的比利时巧克力。我惊讶地看着这一切，他耸耸肩，顽皮地对我眨了眨眼说："你还能期待什么呢？来比利时人的家里，当然要有比利时巧克力。"我拿起一盒Côte D'Or（克特多·金象）品牌的巧克力，这个品牌已有140多年的历史，包装上写着"Noir de Noir"（黑中之黑），意指其可可含量极高。果然，包装背面标注着可可含量高达54%。我又注意到包装左下角贴着一个绿色的"Cocoa Life"标志。总领事介绍说，这是一个全球可可的可持续发展计划，现在大多数比利时巧克力制造商都从与Cocoa Life合作的农民那里采购可可豆。该计划旨在帮助主要的可可生产国提高可持续性可可的种植与生产。他还提到，可可主要生产国的农民面临着资源匮乏、气候变化、种植技术落后以及劳动力廉价等多重挑战。而绿色叶子标志确保可可的获取方式符合生态可持续发展原则，包括可可生产者获得公平工资，且利润在生产者和分销商之间公平分配等。

　　我们边聊边动手制作巧克力慕斯，以下是比利时巧克力慕斯的简单5步制作法：首先，将几块巧克力放入一个耐热的铁碗中，加热至完全融化；然后，在一个干净的碗中加入蛋白和白砂糖，打发至硬性发泡（可以使用电动搅拌器）；接着，将奶油倒入另一个干净的碗中，搅拌至奶油变稠后放入冰箱冷藏；随后，将融化的巧克力倒入打发的蛋白中，轻轻拌匀，再加入奶油继续搅拌均匀；最后，将混合好的巧克力慕斯倒入小杯子中，冷藏2小时即可享用。根据个人喜好，你还可以在慕斯中加入奶油或各种水果，如草莓、蓝莓等，为其增添风味和色彩。

　　在一个闲适的夜晚，何不亲手制作一杯巧克力慕斯，伴着悠扬的爵士乐，享受巧克力与奶油交融的丝滑口感与馥郁芳香？

比利时巧克力慕斯
Belgian Chocolate Mousse

Indulge in a Cup of Belgian Chocolate Mousse

When you think of Belgium, chocolate undoubtedly comes to mind. Belgian chocolate is renowned worldwide for its exquisite taste and exceptional quality, which is deeply rooted in its meticulous craftsmanship and the use of the finest ingredients. For instance, the finer the cocoa powder particles in a chocolate bar, the smoother and more velvety the texture, allowing it to melt effortlessly in your mouth. Belgian chocolatiers have mastered the art of refining cocoa powder to an incredibly fine degree, boasting a particle size smaller than 20 microns—often as fine as 18 microns—far surpassing the standard 30 to 35 microns used in the United States. Furthermore, Belgian regulations stipulate that the cocoa content in chocolate must exceed 35%, significantly higher than the 25% required by the European Union. This higher cocoa content reduces the need for sugar and milk fats, resulting in a richer, more aromatic chocolate. To ensure purity, Belgian chocolatiers insist on using only pure cocoa butter, firmly rejecting any use of vegetable oils.

I had the pleasure to be invited to the residence of Wim Peeters, Consul General of Belgium in Guangzhou. During our discussion on sustainable cuisine, I asked him to recommend a few simple Belgian dishes, and the first thing that came to his mind was Belgian chocolate mousse. "Perfect!" I exclaimed. After all, chocolate, waffles, and beer are some of Belgium's most famous exports.

It's said that in this small kingdom of chocolate of just over 11 million in habitants, the Belgians have an almost obsessive love for chocolate, consuming it nearly every day. Their per capita chocolate consumption is 10.7 kilograms per year, making them the second highest in Europe, just behind the Swiss at 11.6 kilograms. In this "Kingdom of Chocolate", there are over

350 chocolate manufacturers, more than 2 000 chocolate shops, and a dazzling array of over 3 000 different chocolate products, leading to fierce competition and the creation of some of the world's finest chocolate brands.

The Consul General explained that with imported Belgian chocolate, you can easily make a delectable chocolate mousse in just five simple steps. Eager to learn, I followed him into the kitchen, where he opened his refrigerator to reveal shelves packed with Belgian chocolate. I looked on in amazement, and with a playful shrug and a wink, he said, "What else would you expect? When people come to a Belgian's home, there must be Belgian chocolate." I picked up a box of Côte D'Or chocolate, a brand with a history spanning over 140 years. The label read "Noir de Noir" (Black of Black), indicating a high cocoa content. Indeed, the back of the packaging confirmed a 54% cocoa content. I also noticed a green Cocoa Life badge on the bottom left corner of the box. The Consul General explained that this is part of a global cocoa sustainability program, where most Belgian chocolatiers now source their cocoa beans from Cocoa Life partner farmers. This program aims to enhance sustainable cocoa farming and production in key cocoa-producing countries, which face numerous challenges, including resource scarcity, climate change, outdated farming techniques, and low-wage labor. The green leaf ensures that the cocoa is harvested in an ecologically sustainable way; that the labourers get a fair wage and the profits are divided fairly between producers and distributors.

As we chatted, we began making the chocolate mousse. Here's the simple five-step method for making Belgian chocolate mousse: First, place a few pieces of chocolate in a heatproof bowl and melt them completely. Next, in a clean bowl, add egg whites and sugar, then whisk until stiff peaks form (you can use an electric mixer for this). Then, pour cream into another clean bowl and whisk until it thickens, then refrigerate. Gently fold the melted chocolate into the whipped egg whites, followed by the whipped cream, until well combined. Finally, pour the chocolate mousse into small cups and refrigerate for two hours before serving. To enhance the flavor and presentation, you can top the mousse with whipped cream or add your favorite fruits, such as strawberries or blueberries.

On a leisurely evening, why not treat yourself to homemade chocolate mousse? Put on some smooth jazz and savor the silky, aromatic blend of cocoa and cream.

向比利时驻穗总领事裴伟岷学习做地道的比利时巧克力慕斯
照片来源：《Lingling探世界》节目
Learning to make authentic Belgian chocolate mousse from Wim Peeters, Consul General of Belgium in Guangzhou
Photo Source: *See the World with Lingling* TV Program

我将好友送给我的礼物——晒干的橘子皮做成的窗帘装饰挂在自家客厅里
I hung in my living room window a curtain decoration made from dried mandarin peels, a gift from a close friend

第 5 章
Chapter 5

美食中的废料利用
Enjoying Delicious Food While Minimizing Food Waste

　　本章专注于食物使用以及减少（回收）这两个环节，与读者探讨如何将多道粤菜美食物尽其用并尽量减少餐厨废弃物的产生。我们亦介绍了一些国外菜肴，期待从中借鉴关乎废料利用和节约粮食、减少浪费的相关经验。其实，在食物以及其废料利用上，广东人向来颇有心得，能将一件食材的各个部位各尽所能，甚至废料利用。最明显的两个例子是各类果皮、瓜皮的再次利用和顺德厨师对于一条大头鱼的食用开发。

　　近年来，习近平总书记一再提倡勤俭节约，杜绝浪费。我们期待借由此章在节约、不浪费以及通过废料利用而在减少碳排放量的路途上与各位共勉，为我们所处的地球减轻负担出一份力。

This chapter focuses on food consumption and reduction (recycling) of waste, exploring with readers how to fully utilize various Cantonese delicacies and minimize kitchen waste. We also introduce some foreign cuisines, hoping to learn from their experiences in waste utilization and food conservation. In fact, Guangdong people have traditionally been adept at utilizing food and their waste materials, able to make full use of various parts of an ingredient, even the waste. The two most obvious examples include the reuse of peels from fruits and melons, and a Shunde chef's comprehensive utilization of a large carp.

We look forward to sharing this chapter with you in the spirit of saving, not wasting, and reducing carbon emissions through better utilization of food, which will reduce the impact on our planet.

陈皮绿豆沙，橘子皮变宝贝

有一年春节前夕，我到新会探访一位好友。当时正值年关，是大红橘丰收的时节。好友的母亲擅长勤俭持家，用亲手晾晒的陈皮为我们泡了一壶普洱茶——茶香四溢，入口便能驱散旅途的疲惫。她家的阳台上，挂满了正在晾晒中的橘子皮，远远望去，像是一串串立体的、橘红色的剪纸装饰，透着浓浓的年味，让人心生欢喜。

好友的母亲告诉我，从她小时候起，家里的橘子皮从来舍不得丢弃，而是用来制作陈皮。她还说，在广东，自制陈皮是一件简单而传统的事情。取来成熟的大红橘，食用果肉后，将剩下的果皮放入沸水中煮沸，去除表面的污垢。然后，将橘子皮晾晒在阳光下，让阳光慢慢蒸发掉橘子皮的水分，使其颜色逐渐变深。当橘子皮完全干燥后，便可将其放入棉布袋装好，放置于通风干燥的地方，随时取用。

陈皮不仅可以用来泡水或冲茶，还能与绿豆和芸香草一同煮成陈皮绿豆沙。这是一道经典的粤式甜品，不仅清肺解热、抗炎消肿，还具有健脾理气、开胃的功效。

煮绿豆沙时，火候和方法都颇有讲究。最传统的做法是：先将1把绿豆用凉水浸泡约2小时，准备好1株新鲜的芸香草；然后，将10克陈皮洗净，用温水泡软后切成细丝备用；接着，将浸泡后的绿豆放入锅中，大火煮沸后转小火，慢慢搅拌，待绿豆脱壳后，用勺子捞起浮在表面的豆壳；最后，加入陈皮丝和芸香草，继续煮1小时，直到绿豆完全煮烂成沙。享用时，可根据个人口味加入适量冰片糖调味。

这道芸香草陈皮绿豆沙，不仅经济实惠、有益健康，还能充分利用橘子皮，堪称低碳生活、绿色饮食的典范。它将传统智慧与现代健康理念完美结合，适量食用能为我们的日常生活增添一份天然的甜蜜与滋养。

陈皮
Dried Mandarin Peels

陈皮绿豆沙
Mung Bean Soup with Mandarin Peel

Mung Bean Soup with Mandarin Peel—Mandarin Peel Leads the Way

One year, just before Chinese New Year, I visited a friend in Xinhui, a region famous for its harvest of Da Hong Ju (aka big red mandarins) just before the festive season. My friend's mother, a frugal and resourceful homemaker, welcomed us with a pot of Pu'er tea with homemade dried mandarin peel, known as chenpi. After just a few sips, our travel fatigue dissipated. On the balcony of her house, there were mandarin peels drying in the air. From a distance, they looked like strings of three-dimensional orange-red Paper Cutting decorations; it filled the air with a festive spirit and made me happy.

My friend's mother told me that from her childhood to adulthood, she would save the peels from local mandarins and carefully preserve them to make homemade chenpi. She also said that making chenpi is a simple yet cherished Guangdong tradition. After eating the mandarins, just boil the remaining peels for a few minutes to remove any dirt from the surface. After boiling, place the peels in the sun and let them dry. Once the peels are completely dry, they can be stored in a cotton bag in a well ventilated area, and then used when needed.

In addition to being used to brew tea, mandarin peel can also be boiled with mung beans and Ruta graveolens, commonly known as common rue, to make mandarin peel mung bean soup. This is a classic Cantonese dessert that can soothe the lungs, strengthen the spleen, reduce inflammation and swelling in the body, promote healthy energy, and acts as a natural appetizer.

There are very particular methods for preparing mung bean soup. The most traditional method is to begin by soaking a handful of mung beans in cold water for about 2 hours. Then, cut off a plant of common rue and wash it. Soak

10 grams of mandarin peel in warm water to soften it, then cut it into shreds and set aside. Place the mung beans in a pot and boil them over high heat. Reduce the heat to low and stir continuously. Once the mung beans have split, use a spoon to remove the shells. Finally, add the shredded mandarin peel and common rue, and simmer for 1 hour until the mung beans turn into a "soil-like form". When enjoying, add a little rock sugar to taste according to your personal preference.

Common rue, mandarin peel, and mung bean are not expensive and are healthy to eat, and the use of the recycled mandarin peel is especially thrifty. Eating plenty of Mandarin Mung Bean Soup is a great way to detoxify and strengthen the body, while enjoying a green diet and living a low-carbon lifestyle.

豉油王煎鱼头，大头鱼无一部位不成佳肴

俗话说："食在广州，厨出凤城。"

有一段时间，因为工作关系，我长住顺德，周末常常流连于大街小巷之中觅食当地特色私房菜。顺德是粤菜的主要发源地。2014年，联合国教科文组织将顺德评定为"世界美食之都"。无论是在广州、香港、澳门，还是在海外华人聚居地的粤菜餐馆中，顺德厨师都备受青睐，尤其以他们对淡水鱼的巧妙运用而驰名海内外。

在寻访顺德美食的旅途中，我经常会吃到当地的淡水鱼，比如"四大家鱼"。其中，大头鱼是最常见的，其他三种家鱼分别是青鱼、草鱼（广东俗称"鲩鱼"）以及鲢鱼。在顺德的普通百姓家中，大头鱼是餐桌上的"常客"，无论是蒸、煎、焖、煮、炸，还是生吃或用作火锅配料，都能见得到它。在酒楼和大排档里，大头鱼更是屡屡成为菜品中的亮点。

顺德厨师可以将一条大头鱼演绎出无数种美味，将它从头吃到尾，丝毫都不能浪费。鱼头、鱼背、鱼腩、鱼骨到鱼皮、鱼尾，无一部位不能成佳肴。顺德民间素有"大鱼头，鲩鱼尾"的讲法，意思是吃鱼头以大头鱼为佳，而吃鱼尾则首选鲩鱼。

一般来说，大鱼头的胶质和腥味都比较重，如果处理不当，鲜味便会大打折扣。顺德厨师通常会先将鱼头腌制，再采用煎焗的方法，使鱼肉在保留鲜美的同时，也更加入味，口感极佳。

在顺德的这段日子里，我有幸品尝到一道豉油王煎鱼头，用的正是大头鱼的鱼头。这道菜的做法简便，且价格实惠。"食过返寻味"（指对于一些味道极佳的菜

看，吃过之后常常牵挂，想要再次品尝一番）之后，我特意向餐馆老板申请，进入后厨观摩学习，最终得以掌握这道菜的烹饪精髓。

那么，如何在家中制作一碟美味的煎焗鱼头呢？关键在于要有好的生抽。一般在超市购买的生抽味道偏咸且略带腥味，直接使用容易掩盖鱼肉本身的鲜甜，得不偿失。为了调制出更佳的风味，这家餐馆的生抽是经过特别调制的——将生抽用清水稀释，加入白砂糖、西芹、冬菇和香菜等材料调味。如果在家制作，为了方便起见，可以使用蒸鱼酱油，这种酱油咸淡适中，还带有一丝鲜甜。如果仍觉得过咸，可以加些清水稀释，再滴入几滴美极鲜酱油和鱼露调味。

当酱油调料准备好后，接下来便是挑选一条新鲜的大头鱼鱼头，去鳃后切成4～6块。用盐和生粉涂抹均匀，加入少许调好的酱油料，用手抓匀后腌制15分钟。大火热锅，加入去皮姜片、红葱头、去皮蒜粒等爆香，再将鱼头用小火煎至金黄，翻炒后盛起。然后，在一个干净的瓦煲中放油，加入蒜头、红葱头和姜片煎至焦黄，将鱼头放回瓦煲中，加入特制酱油调料和胡椒粉，表面铺上葱段和小红辣椒段，盖上煲盖焖焗10分钟即可。

还记得在我小升初那年，连续好几个星期，家里的餐桌上几乎每天都有鱼头，担任厨师的奶奶总是说："食多滴鱼头啦，聪明啊！"（"多吃一点鱼头吧，这样可以变得更聪明"的粤语表述。）当时，我只是觉得金灿灿的煎焗鱼头味道肥美，肉质嫩滑，因此吃了不少。后来，通过一番努力，我考上了省里的重点中学。虽然尚不得而知这是否与吃了鱼头而变得聪明有关，但在广东，确实流行有"吃鱼头变聪明"的说法。

194

豉油王煎焗鱼头制作中
Bighead Carp with Soy Sauce in the making

豉油王煎鱼头
Bighead Carp with Soy Sauce

Bighead Carp with Soy Sauce— Making the Entire Fish Delicious

There is a local saying which directly translates, "Guangzhou is the paradise of food, while Shunde is the cradle of chefs." What this actually means is that if food in Guangzhou is impressing you, you know the culinary skills must have come from the Shunde region.

In 2023, I lived in Shunde for an extended period of time. On the weekends, I explored the local streets and alleys looking for good food and especially delicious home-made cuisine (私房菜). Shunde is a birthplace of many types of Cantonese cuisine and was named a "City of Gastronomy" by UNESCO in 2014. Apparently, Cantonese restaurants in Guangzhou, Hong Kong, Macao and overseas Chinatowns like to hire Shunde chefs, who, among other things, are good at cooking freshwater fish.

In my culinary adventures through Shunde, I saw a lot of freshwater fish, especially the bighead carp(大头鱼)(The other common freshwater fish in Shunde are the black carp, the grass carp and the silver carp). In ordinary homes, this dish is often seen on the dinner table: steamed, fried, braised, boiled, eaten raw or in a hot pot; you name it. It was also readily available it in restaurants and food stalls.

A bighead carp, in the hands of a Shunde chef, can be made into countless different dishes. You can eat it from head to tail without wasting anything. The fish head, fish back, fish belly, fish bone, fish skin, fish tail—every part can become a delicacy—which is simply amazing. Shunde people have a saying "bighead carp head, grass carp tail", which means that grass carp tail is the most delicious fish tail, and the best fish head is from the bighead carp.

Generally speaking, the colloid and fishy smell of the fish head can be quite pungent. If not handled well, the flavor could be distasteful and overpowering. The method of first marinating, and then light frying, can eliminate the fishy smell, preserve the natural taste of the fish, and results in a very tasty dish.

In my food hunt, I came upon a "pan-fried fish head with soy sauce", dish which featured the head of a bighead carp. After being captivated by its flavor, I asked for permission to observe the chef prepare this dish in the kitchen and am now sharing with you the recipe and secrets for doing it well.

So, how can we make a delicious pan-fried fish head with soy sauce at home? The key is to have good soy sauce. Generally speaking, the soy sauce bought in the supermarket tastes salty and sometimes even a little fishy. If it were used directly, the fresh sweetness of the fish would likely be covered up. But the soy sauce used in this restaurant is homemade. The chef dilutes the supermarket-bought soy sauce with water, and then blends in white sugar, celery, shiitake mushrooms and coriander to taste. If you want to save time at home, it is recommended to use a "steamed fish soy sauce" variety, because it is moderately salty and has a slightly sweet flavor. Another option is to add a little water to dilute regular soy sauce, and then add a few drops of Maggi fresh soy sauce and fish sauce to taste.

Now you have your homemade sauce and you have a fresh bighead carp head at hand with the gills removed. Cut the fish head into 4-6 pieces, smear it evenly with salt and corn flour, add a little of the homemade sauce, and marinate it for 15 minutes. Heat up oil in a pan for stir-frying, then add the peeled ginger, red onion and peeled garlic and pan fry the fish head slowly until golden. Then, put oil in a clean clay pot and fry garlic, red onion and ginger until golden brown. Put the fish head to the pot, add some special soy sauce seasoning and pepper, spread the onion and red peppers on the surface, cover the pot, and braise for 10 minutes.

Once upon a time, when I was in the final year of primary school, we had a fish head with soy sauce dish on our dinner table almost every day for several weeks. Grandma, who was the cook at home, always said:"Eat more fish heads, as it makes one smarter." At that time, I only thought that the golden fried fish head tasted great, and that the meat was tender and juicy, so I ate a lot. Later, after some effort, I was admitted to a highly selective provincial junior high school. Whether it had anything to do with my having eaten so much fish head will always be a mystery (wink), but in Guangdong, the saying "eating fish head makes one smarter" has been popular for centuries.

虾壳汤泡饭与虾壳的二次利用

我们在制作甜酸咕噜虾时，常常会剩下许多虾壳。很多人会把这些虾壳拿去喂猫，虽然这样也算物尽其用，但总觉得有点可惜。或许，有人会将其制成虾酱或其他加工产品，但这需要专业的技术和设备。其实，虾壳中富含丰富的钙质，非常有营养，但由于其质地坚硬，不宜直接食用，否则将不利于肠胃的消化，所以最好经过加工后再食用。

如果你喜欢虾壳的鲜味，并且想充分利用它，那么不妨用虾壳来做一碗虾壳汤泡饭。这是一道简单易做又开胃爽口的菜肴。

制作虾壳汤泡饭的材料非常简单：准备一些芥蓝粒、香菜段、葱花、炒米、番茄粒，再留一些之前做虾时的虾肉碎。将虾壳放入热油锅中煎至微黄，加入热水煮出鲜美的虾壳汤，接着放入调味粉和所有配料，再加入隔夜的米饭稍微煮一下即可。这道菜品采用少油、少盐的烹饪方式，让虾壳的天然鲜味得以充分展现。

我有一位暖男朋友，很会哄老婆开心。他常常买一些新鲜的基围虾回来，为妻子亲自下厨烹饪，两人还会在烛光下享受美酒佳肴。对于剩下的虾壳，他从不浪费，而是留着第二天与妻子一起制作有机肥。他告诉我，制作有机肥的最简单方法是先将煮熟的虾壳捣碎，放进一个瓶子里，让其充分腐烂后，连壳带水倒入花盆底部，但需注意避免接触到花根。他说，虾壳是一种非常优质的磷肥，用来给兰花施肥，兰花年年都能开得绚烂。

剩余虾壳
Leftover Shrimp Shells

炒熟的虾壳
Stir-Fried Shrimp Shells

虾壳汤泡饭
Shrimp Shell Soup with Rice

Shrimp Shell Soup with Rice— A Tasty Way to Make Use of Shrimp Shells

When we make sweet and sour "gulu" prawns (which I talked about in the previous chapter), we leave behind a lot of shrimp shells. What can we do with them? one may wonder. Some people use shrimp shells to feed cats, but in fact, the shells of shrimp are edible for people as well, which are rich in calcium. However, although shrimp shells contain rich nutrients, they are not suitable for direct consumption because they are hard on digestion. Therefore, it is best to consume them after processing into shrimp paste or another product, but this requires professional technology to do properly.

I think if you like the freshness of shrimp shells and want to maximize their use, you can make shrimp shell soup and rice. This is an extremely simple, easy to cook, and appetizing dish.

First, prepare some mustard greens, coriander segments, scallions, fried rice, diced tomatoes, and leave some shrimp meat to make minced shrimp when making sweet and sour "gulu" prawns. Fry the shrimp shells in a frying pan, pour in hot water as a base for the shrimp soup, then add some seasoning powder and let all the ingredients simmer overnight. This dish follows the principle of less oil and salt, allowing the fresh taste of the shrimp shell to come through effortlessly.

I have a warm and considerate friend who is excellent at making his wife happy. He often buys fresh shrimps and cooks these for his wife at home, and the couple would enjoy them with wine during a candlelight dinner. He will save leftover shrimp shells to make organic fertilizer for his wife the next day. He said that the most convenient way to make organic fertilizer is to first mash the cooked shells and put them into a bottle, let them rot as much as possible, then pour the shells and water into the bottom of the flower pot, but do not let them touch the root and stem of the flower. He said that shrimp shells are a very good phosphate fertilizer and can be used to fertilize orchids, keeping them in bloom year after year.

节约但不简约，日式早餐瘦身有妙招

刺身、三文鱼手卷或天妇罗等日式美食早已广为人知，但在我心中，最令人印象深刻的还是那丰富多样、精致优雅的日式早餐。

记得初次品尝地道的日式早餐，是在一次日本旅行中。清晨走进酒店餐厅，眼前的自助餐台上整齐摆放着各式精美小菜：纳豆、味噌汤、烤鱼、撒上葱花的米饭、金针菇、玉子烧、菠菜、酱腌白萝卜、小番茄、哈密瓜，种类繁多，琳琅满目。那一刻，我不禁感叹，哇，原来日式早餐竟如此丰盛，食材的丰富程度令人惊叹。然而，令人费解的是，日本街头的行人却大多身形纤瘦。究竟是什么让他们能够在享受美食的同时，保持如此健康的身材呢？

后来，我在早餐厅细心观察，发现日本人用餐时，每次只会精挑细选几样食物，分别盛装在一个个精致的小碟子、小碗里。而那些碗碟的尺寸之小几乎令人难以置信，仿佛只需两三口就能将其中的食物吃完。这或许就是日本人保持苗条身材的秘诀之一——虽然食物种类繁多，但每样都只是浅尝即止。另外，我很少见到有人将餐盘装得满满的，更少见到因自助餐而过度进食、浪费食物的现象。节约食物、文明用餐的理念在这里得到了淋漓尽致的体现，是值得学习的好习惯。

之前有幸受邀前往日本驻广州总领事馆官邸做客，并再次享用了一顿地道传统的日式早餐。当日，总领事官邸料理人梅泽先生亲自为大家下厨（梅泽先生曾经于东京帝国饭店出任大厨，资历深厚，已在广州生活10来年）。

这份早餐包括了盐烤三文鱼搭配甜玉子烧、洋葱纳豆、海苔烧金针菇、时令蔬菜以及白米饭搭配味噌汤。除了盛米饭的碗和装三文鱼的盘子外，所有餐具都小巧精致，颇具日式早餐的典型风格——每个餐具直径不过10厘米。色彩缤纷的陶瓷餐具形态各异，有的形如花瓣，有的像一片叶子，赏心悦目。这些小碟子、小碗整齐排列在托盘中，每人一份。酱油在这顿早餐中扮演了重要角色——纳豆、海苔烧金针菇和味噌汤中都使用了酱油；而三文鱼和玉子烧则以清淡为主。吃一口进嘴里，除了感受到酱油的香气，更能品味到食材本身的鲜美。

我想起现今流行的一种说法："精致的摆盘是日本料理的特色，而西洋料理的精髓在于其香料和草药的芳香，至于中国料理的卓越之处则在于其烹饪技艺的高超。"（引自石毛直道《日料的故事》），我边看边吃，更多的是用眼睛在享受——碟中矗立的甜玉子有点像是日本庭院中的山石，盐烤三文鱼上的油光闪闪让我联想到庭院池水的波光粼粼，而绿色的紫苏叶则象征着庭院中的树木。高低错落的立体摆盘，左右不对称，仿佛在展现日本庭院美学的精髓。

其实，在家也可以动手尝试制作日式早餐。当然，家庭用餐的首要考虑是营养和饱腹感，而摆盘美学则视情况而定。传统的日式早餐通常包括一块烤三文鱼、一碗米饭配拉丝纳豆、一到两款时令水煮蔬菜、玉子烧，再搭配味噌汤和一杯热抹茶。

如果想要口感更加丰富，可以用清酒轻腌三文鱼，煎熟后撒上一点白芝麻。若想省时省力，玉子烧可以用煎蛋替代，味噌汤也可选用超市购买的半成品。目前，在广州的许多超市都可以买到日本味噌汤，只需加点葱花碎即可提鲜。这份日式早餐低盐、低钠、低糖、低脂肪，既能提供满满的能量，又不会带来多余的热量。

俗话说，三分练七分吃。我想，日本大多数人身材苗条，一定是因为他们有着注重食材的新鲜和营养，所吃分量不大，更讲究重质、重观、不重量的饮食习惯。

日式早餐
Japanese breakfast

日式早餐中的三文鱼
Salmon from the Japanese Breakfast Set

Japanese Breakfast—An Economical, Simple (But Not Shabby) Meal, That Promotes Healthy Living

Lunch and dinner foods like sashimi, salmon rolls, or tempura may be the most familiar Japanese foods, but my personal favorite is those eaten at a traditional Japanese breakfast.

I vividly recall my first encounter with an authentic Japanese breakfast in Japan. As I walked into the hotel dining room that morning, I was greeted by a beautifully arranged buffet spread with all kinds of foods—natto, miso soup, roasted fish, rice sprinkled with chopped green onions, needle mushrooms, tamagoyaki grilled eggs, spinach, white radishes, cherry tomatoes, cantaloupe—tempting me by their appearance. I couldn't help but think, wow, Japanese people eat so much food. But most of the Japanese people I saw were slim. What is the reason for this?

Later, I carefully observed the Japanese guests in the breakfast room. They would usually carefully select three to five items and put each in an exquisite small plate or bowl. The bowls and plates were so small that you could eat the food held there in one to three bites. I thought perhaps I had found the reason, one at least, as to how Japanese people can keep slim; they eat a lot of different foods but in small portion. In addition, I rarely saw Japanese people filling up their plates and letting a lot of food go to waste.

I was recently invited to Consulate General of Japan in Guangzhou, where I once again experienced an authentic Japanese breakfast. The meal was prepared by the residence's chef, Mr. Umezawa, who worked formerly at the

Imperial Hotel in Tokyo and had extensive experience. He had been living in Guangzhou for over 10 years.

The breakfast featured salt-grilled salmon paired with sweet tamagoyaki, onion-topped natto, needle mushrooms wrapped in seaweed, seasonal vegetables, and steamed white rice served with miso soup. Except for the rice bowl and the plate for salmon, all of the plates were small, which is typical Japanese breakfast ware; each piece no more than 10 centimeters in diameter. The colorful tableware came in various designs, some shaped like flowers or leaves, and all pleasing to the eye. They were laid out on a tray, one tray per person. Soy sauce was the main condiment—used in the natto, seaweed braised needle mushroom, and miso soup. The seasoning used for salmon and grilled eggs was minimal and very limited. The experience of eating was about experiencing the natural flavor of the ingredients, with a rich taste of the sauce.

I found a quote from the book *Stories of Japanese Cuisine*, which roughly translated states that "exquisite food arrangement is a core element of Japanese cuisine, while western cuisine is characterized by the distinct spices and herbs used, and the excellence of Chinese cuisine lies in the culinary technique". I watched and ate, and enjoyed the arrangement in front of me. The dish reminded me of a Japanese garden with the sweet tamagoyaki standing on the plate like mountains and rocks, the oil on the baked salmon sparkling like sunlight on a pond, and the green perilla like leaves on trees. This three-dimensional pattern with highs and lows, asymmetric yet balanced, is very much in line with the aesthetic of a Japanese Garden.

You could always try making authentic Japanese breakfast at home. Of course, when dining at home, the first thing to consider is nutrition and the pleasure of eating a satisfying meal, and the aesthetics of plate arrangement can come second. The traditional Japanese breakfast usually includes a piece of baked salmon, a bowl of rice, one or two boiled seasonal vegetables, grilled sweet egg rolls, miso soup and a cup of hot matcha tea. (In fact, before the 1950s, the typical breakfast menu in Tokyo was rice, miso soup, natto and pickles, perhaps because Japanese culture does not value spending too much time cooking.)

If you want the best taste, marinate the salmon with sake lightly, and sprinkle some white sesame seed on top after pan-frying. If you want to save time and trouble, grilled egg rolls can be replaced with fried eggs, and you can buy instant miso from the supermarket in advance. Nowadays, many supermarkets in Guangzhou sell Japanese miso soup mixes. But remember to have some chopped green onion to help freshen it up. This Japanese breakfast is low in salt, sodium, sugar and fat. Eating it will bring you energy for the whole morning and not excess calories.

There is a saying in body building and weight management, success comes from "30% exercise and 70% diet". It is rare to see overweight people in Japan; I think this definitely has something to do with small portions of food and an emphasis on freshness and nutrition. Simply put, there is a focus on quality and value over quantity.

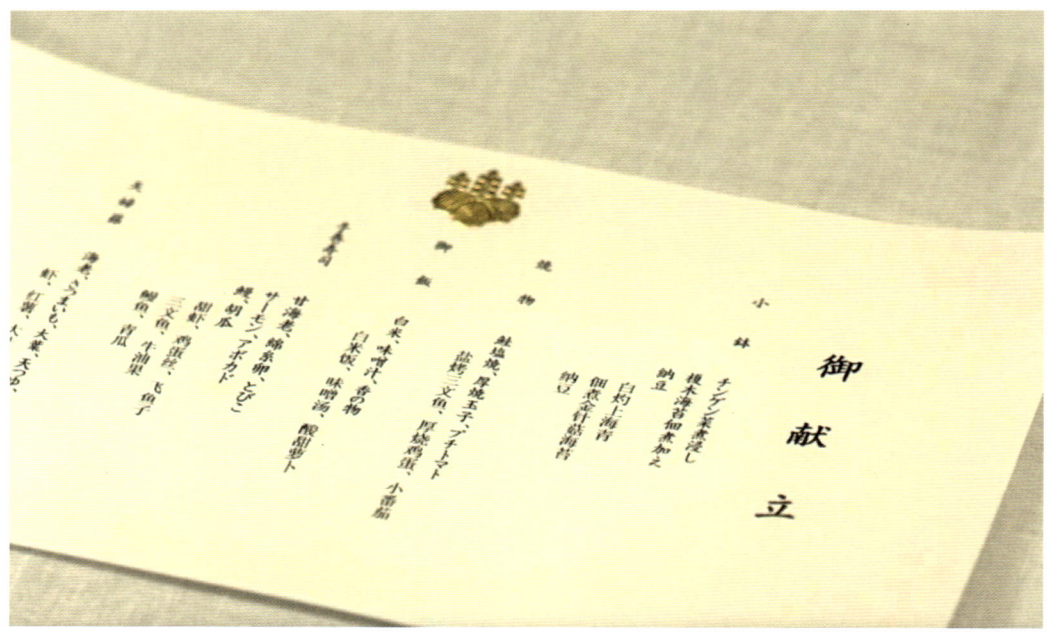

日本驻广州总领事馆的早餐招待会餐单
Breakfast reception menu from the Japanese Consulate General in Guangzhou

意大利土豆鸟基，让亲子关系更加亲密

在意大利，流传着这样一句谚语："生活是魔法与意大利面的结合。"这句话形象地表达了意大利人对于面食的钟爱。多年来，我尝遍了常见的各种意大利面：细面（spaghetti），千层面（lasagna），笔管面（penne），宽面（tagliatelle），宽条面（fettuccine）。然而，最令我念念不忘的，还是那款融入了土豆的意大利鸟基面，那种独特口感令人回味。

鸟基是意大利独具特色的面食，由土豆与面粉混合而成，煮熟后再加入精心调制的番茄酱汁。之所以叫鸟基，并无特殊含义，不过是意大利文的音译。

在意大利，鸟基几乎可以算作是一道全民美食，当地人经常说"Giovedi gnocchi"，意思是"今天是周四，让我们吃鸟基吧"。在周四，人们会选择吃鸟基，在不知道该吃什么的时候会做鸟基；外出就餐时，也经常点鸟基。虽然意大利各个大区对鸟基的制作方法略有不同，但传统的基础版鸟基一定包括4种原材料：低筋面粉、鸡蛋、土豆、番茄酱汁。做法是先将土豆用开水煮30分钟至熟，之后再用微波炉加热以蒸发掉多余的水分；接着，用勺子将其压成泥状。将面粉、鸡蛋和土豆泥混合后充分揉好，搓成长条状，再将其切成直径2厘米左右的小块备用。将鸟基面团静置20分钟后再进行烹煮——取大容量锅，装水加盐煮沸，要吃的时候，将鸟基面团放入沸水中煮熟，然后捞出放入番茄酱汁，简便快捷。

有一次，我在意大利住在朋友家，看到她的母亲从清晨便开始在厨房忙碌，先是在上午准备传统的意大利番茄酱，然后下午制作鸟基。我看她制作鸟基，觉得非常有趣。她将面团揉好后，切成一个个小小的面疙瘩，然后用叉子轻轻压出纹路，有的竖、有的横，纹路或深或浅，不仅美味，还很精致。当时我就在想：在这个节奏飞快的现代社会，如何在尽可能少的时间内完成更多的事情，成为我们的思考课题。如果妈妈们花1小时做饭，再花1小时陪伴孩子用餐，那么纯粹陪伴孩子的时间只有1小时。但如果将烹饪过程变成亲子活动，比如让孩子一起参与鸟基的制作，那么亲子相处的时间相当于延长到了2小时——不仅烹饪了美食，还增进了亲子之间的感情，真正实现了"1加1大于2"的效果。

　　如果你在广州想要品尝地道的意大利鸟基，可以前往OGGI意大利餐厅。其前总经理安德鲁是一位意大利人，他将家乡的鸟基引入了广州。安德鲁说，传统的鸟基通常搭配番茄酱作为汤底，但在意大利菜中，番茄酱是最难调制的酱汁之一，因为番茄的酸甜程度常常受当天番茄采购质量的影响，很难把握。为了让广州的食客更易接受这道菜，安德鲁对鸟基进行了改良和创新，改用大虾熬制的汤汁作为汤底，使口感更加鲜美。另外，他还沿用了意大利部分地区在鸟基面团里加入帕马森干酪的做法，让它吃起来更加筋道和香口。

意大利鸟基面团
Italian Gnocchi Dough

意大利鸟基
Italian Gnocchi

Italian Gnocchi—Bringing Families Closer

In Italy, there's a saying, "Life is a blend of magic and pasta." (by Federico Fellini) Italy is famous for having a passion for pasta. Over the years, I have tasted various types of Italian pasta, the most common ones being spaghetti, lasagna, penne, tagliatelle, and fettuccine, the list goes on and on. Yet my favorite is the gnocchi pasta made from flour and potato, which left a lasting impression after just one taste.

Gnocchi is a specialty Italian pasta dish made by combining potatoes with flour, then boiling the mixture. After being cooked, a tomato soup base is added and it is ready to serve.

Gnocchi is loved throughout Italy. Local people often say "Giovedi gnocchi", which means "Today's Thursday, let's have gnocchi". Many local households prepare gnocchi on Thursdays, and on any given day when they can't figure out what to eat, they also make gnocchi. Gnocchi is also a favorite dish when dinning out. Although gnocchi recipes vary across different regions of the country, an easy traditional Italian homemade potato gnocchi is almost always consisted of 4 ingredients—flour, eggs, potatoes, and tomato sauce. To prepare, first boil the potatoes for about 30 minutes, then use a microwave to heat and evaporate excess water, mash the cooked potatoes with a spoon. Mix the flour with eggs and combine them with your fingers to form a soft dough (it should not stick to your fingers). On a lightly floured surface, cut small portions of dough, roll to form ropes, and then cut into 3/4 inch (2 centimeters) pieces. Let the gnocchi set for 20 minutes before cooking. Boil the gnocchi in a large pot of water with some added salt. The gnocchi are ready when they float to the top. Drain and toss with sauce of your choice.

Once while staying at an Italian friend's house in Italy, I saw her mama busy in the kitchen. She began early in the morning, making traditional Italian tomato sauce, and in the afternoon made gnocchi. Watching her making the gnocchi was especially interesting to me. She pressed each flour rope gnocchi with a fork to make different lines and patterns, some vertical and some horizontal, some deep and others shallow. Not only was the food she prepared delicious, it was also lovely. I was thinking at the time that in today's fast-paced society, we're always trying to do more in less time. If a parent spends one hour cooking alone and then one hour eating with the family, then the family time is only one hour. But if preparing the dish can be turned into a shared family activity, for example, allowing children to help press patterns on the gnocchi, then the family time is extended to two hours.

It's a bit like one plus one equals more than two.

If you're in Guangzhou and crave authentic Italian gnocchi, you should visit OGGI Italian Restaurant. Andrew, the former general manager, introduced the gnocchi recipe from his hometown to Guangzhou a few years ago. His traditional gnocchi recipe used tomato sauce as the base, and Andrew explained why tomato sauce was considered one of the most difficult sauces to make. This is because the sweetness and acidity of a tomato sauce is greatly affected by the tomatoes purchased on any given day. So Andrew localized the dish by incorporating seafood elements loved by Cantonese locals, for instance, using a broth made from prawns instead of the traditional tomato. To finish the dish Andrew followed the traditional practice of topping with fresh Parmesan cheese.

德国红酒炖牛肉，对过夜美食 say yes！

德国位于欧洲中部，处于高纬度地带，夏季温和，平均气温在30℃以下；冬季极为寒冷，尤其是中部和南部地区。德国的大都市慕尼黑，历史上曾记录过零下27℃的低温，堪称酷寒。在这样的气候条件下，当地人偏爱饮用啤酒和红酒，食用牛肉和羊肉来祛寒暖身。使用小牛肉块和前一天喝剩下的红酒制作的红酒炖牛肉（Goulash），便是德国一道经典的家常菜肴。（值得一提的是，"古拉什"红酒炖牛肉最早起源于匈牙利，后来传播到德国，并在当地广受欢迎。）在德国有句俗语："周五吃鱼，周一至周四吃肉。"由此可见，这道炖牛肉不仅是一道美味，更是人们在忙碌之后恢复精力的能量之源。

制作红酒炖牛肉所需的主要食材包括切成小块的牛里脊肉，浓稠的番茄酱，约150毫升的红酒（许多人会用前一天喝剩下的），2瓣大蒜，两三片香草叶，切成条的小红辣椒，切片的洋葱。首先，将牛肉块放入沸水中焯水去腥；用橄榄油热锅后，放入蒜片和洋葱翻炒，再将焯过水的牛肉块放入锅中翻炒至五成熟，然后加入浓缩番茄酱和剩下的红酒，快速翻炒至牛肉呈现焦黄色。将锅中的所有食材转移至另一口炖锅中，加入清水、香草叶、小红辣椒，转小火慢炖1.5~2小时即可。

需要注意的是，红酒不宜使用开瓶超过3天的，最好在2天内用完，因为存放过久的红酒容易变酸，除非你希望炖出的牛肉带有一丝酸味。如果喜欢更辣的口感，除了加入小红辣椒，还可以调入一些塔巴斯哥辣酱（Tabasco），刺激一下味蕾，带来一种冬日午后从昏昏欲睡中被辣味唤醒的感觉。不过，使用时需谨慎，毕竟这是一种相当刺激的调料。

　　炖好的牛肉可以搭配米饭或面包享用。一位居住在广州的德国外交官朋友告诉我："在德国，这道菜是许多家庭主妇为丈夫精心准备的晚餐，满满的热量和爱意融入其中。"想象一下，劳累了一天的丈夫回到家中，推开门便闻到满屋飘香的炖牛肉味道，夹杂着红酒的芬芳，令人食欲大开，同时也更增添了丈夫对妻子的感激与爱意。他还说，"这道菜不仅可以将剩下的红酒物尽其用，还能保存几天，随着时间的推移，味道更加浓郁。"是呀，经过一夜的浸泡，酱汁更加入味，肉质更加鲜美。第二天上班前，将炖好的牛肉装入午餐盒中，配上一些沙拉和黑面包带到公司，作为午餐享用，既省心省时，还能享受美味佳肴的陪伴。

德国红酒炖牛肉
German Goulash

Goulash—One Way German Cuisine Uses Leftovers

Germany is located in central Europe, at a high latitude. In summer, the weather is relatively mild, with an average temperature below 30°C(86°F). But in the winter it can get pretty cold, especially in the Central and Southern regions. The lowest recorded temperature in the German city of Munich is -27°C(-16.6°F). Germans food culture is famous for beer and wine, and for making beef and lamb stews to stay warm. Goulash, a classic home-made dish, is cooked by stewing cubed veal with red wine sauce, and it is popular to use leftover red wine to make goulash. (It is worth mentioning that Goulash originated in Hungary and later spread to Germany, where it gained widespread popularity.) There is a local saying in Germany, "eat fish on Friday night and red meat on Monday through Thursday nights." It should come as no surprise, then, that Goulash is one of the several dishes to keep people warm and well-fed during the busy work week.

Goulash typically includes the following ingredients:beef tenderloin cut into small pieces, tomato sauce, red wine, garlic, 2-3 vanilla leaves, small slices of red chili pepper, and julienne onions. To prepare, blanch the beef cubes in hot water. Next stir fry the beef with the garlic slices and onions in oil until they are about half cooked. Then add the concentrated tomato sauce and red wine. Quickly stir fry the beef cubes until they are brown. Finally transfer all the ingredients to a stew pot and add water, vanilla leaves, the red chili pepper. Simmer over low heat for 1.5 to 2 hours.

When using leftover wine for this dish, it is best to use within 2 days of opening, or the wine might taste more like sour vinegar, and generally speaking red wine is not recommended for consumption if it has been opened for more than 3 days. If you want more of a kick, in addition to adding small red chili peppers, you can also add a little Tabasco sauce. But be careful as it can quickly get much spicier.

Goulash is typically eaten with rice or bread. A German diplomat friend once told me that in Germany, this dish is said to carry the love of a romantic partner, full of energy and affection, and is a common dish made at the home. Imagine returning home from a busy day to the aroma of goulash red wine; the love of a partner is felt without the need for words. This dish is not only a great way to use leftover red wine, but it is also a dish that can be eaten on for several days. Many people feel that after a night or two stewing in the sauce, the meat in the leftover goulash becomes even more tender and tasty. In Germany, it's common to pack this dish in a lunch box to have at work the next day, along with a little salad and bread, for a delicious and convenient lunch.

赠送《中国粤菜故事》给德国前驻穗总领事冯马汀，与其交流德中两地饮食文化之异同
Presenting *The Story of Cantonese Cuisine* to Martin Fleischer, former Consul General of Germany in Guangzhou. We also had a conversation on culinary traditions of Germany and China

善用边角余料的沙克舒卡

以色列位于地中海的东南方向，坐落于亚、非、欧三大洲的交会处。这里既融合了欧、亚、非三地的风情，又孕育了世界三大宗教——基督教、犹太教和伊斯兰教。如果要挑选一道能够代表以色列的菜肴，我会毫不犹豫地选择沙克舒卡。这道以色列的传统早餐色彩斑斓，仿佛是一幅流动的马赛克拼接画，每次烹饪出来的都有些不同。它象征着以色列现存的各种文化，这些文化在历史的长河中并行发展，彼此交融碰撞。不同色彩的文化拼接在一起时，既有撞色的冲突，也有柔和的渐变。它们的每次交织，都在画板上编织出奇幻且让人着迷的图案。

然而，这样一道外表美丽的菜肴，却是用极其普通，甚至是前一天剩下的食材制成的。沙克舒卡之所以在以色列广受欢迎，正是因为它不需要特意准备食材。在20世纪五六十年代，犹太移民回归潮达到高峰期，许多人经济拮据，为了填饱一家人的肚子，常常需要精打细算。而制作沙克舒卡，只需将前一天剩下的各种蔬菜边角料，如青椒、黄椒、小红辣椒、香菜、樱桃番茄以及蒜片等炒在一起，如果喜欢，还可以加入一些罐装浓缩番茄酱，再打上2个荷包蛋，一道美味的佳肴便呈现在眼前。我想，沙克舒卡之所以能够成为以色列的国民美食并流传至今，很大程度上与犹太人崇尚节俭、质朴的生活理念有关。

"沙克舒卡"这个名字源于非洲当地的一种方言，意思是"将所有东西混在一起"。这道菜起源于非洲东北部的利比亚和突尼斯地区，后来由居住在那里的犹太移民带回以色列。因此，沙克舒卡还有一个有趣的中文名字"北非蛋"。

我第一次品尝到这道菜是在以色列驻广州总领事高文先生的家中。那天，总领事先生与夫人亲自下厨，准备了这道佳肴。首先，他在一个大平底锅中倒入橄榄油并加热，加入切碎的番茄、大蒜、中辣度的青椒、少量小红辣椒段以及番茄酱，不断搅拌直到番茄里的大部分水分蒸发。那红红的番茄酱看上去就像浓郁的红咖喱，只是番茄酱更加健康。接着，高文先生在番茄酱上打了2个荷包蛋，再加盐调味，改用文火，盖上盖子，焖煮20分钟直到蛋黄完全煮熟。让我感到惊喜的是，高文先生在刚出锅的沙克舒卡中加入了我们广东人常用的香菜。他告诉我，他非常喜欢使用香菜，不仅在做沙克舒卡时，在制作其他以色列菜肴，如胡姆斯时，也会用到香菜。其实，香菜的原产地最早就是地中海沿岸及中亚地区。

　　当然，在世界各地的犹太人聚居区，会有不同版本的沙克舒卡。有些人会加入更多的青红辣椒来增加辣味，有些人则会撒上一些孜然粉——对以色列人来说，孜然粉就像我们常用的白胡椒粉一样普遍。有的人还会搭配一些菲达奶酪和香肠，从素食清淡到咸辣重口味等各种口味，全凭个人喜好调整。

　　沙克舒卡让我联想到家乡的广东炒饭。它也源于一些并不富裕的家庭，为了节省时间和开支，将剩下的米饭和一些蔬菜、鸡蛋以及腊肉炒在一起，与沙克舒卡有着异曲同工之妙。

可用于制作沙克舒卡的边角余料
Kitchen Scraps for Shakshuka

以色列沙克舒卡
Israeli Shakshuka

Shakshuka—A Great Dish to Use Leftovers from the Israeli Kitchen

Israel is located in the Southeastern corner of the Mediterranean, at the crossroads of Asia, Africa, and Europe, and converging with the customs of those very diverse cultures. It is also a place from which three major religions of the world were born: Christianity, Judaism, and Islam. If I had to choose a dish that represents Israel, I would choose an Israeli breakfast dish called Shakshuka. The vibrant colors of Shakshuka are like a moving mosaic; each time this dish is cooked it looks slightly different. Shakshuka is also representative of the various cultures which co-exist in Israel, developing side by side and influencing one another in the process. When these diverse cultures come together, there are contrasting colors and many opportunities for fusion, like modern art weaving fantastic and captivating patterns on a canvas.

Interestingly, this dish, which is so aesthetically pleasing, is made with ordinary ingredients, and often leftovers from the previous day. I think that perhaps this was a key reason that Shakshuka first became so popular in Israel. In the 1950s and 1960s, when the wave of immigrants to Israel reached its apogee, budgets were tight, and people needed to be frugal and resourceful. To make a Shakshuka, one could stir fry leftover vegetables from the previous day, including green and yellow peppers, chili peppers, coriander, cherry tomatoes, and garlic slices; add a little canned concentrated tomato sauce, a couple of poached eggs, and a delicious dish was born! Shakshuka continues to be an Israeli national delicacy to this day and is representative of Israeli cultural values of frugality and simplicity.

I found it interesting to learn that "Shakshuka" is a word from an African dialect meaning "mixing everything together", and the dish originated from the Libya/Tunisia region of northeastern Africa.

It was introduced to Israel by Jewish immigrants from those regions. Shakshuka also has an interesting Chinese name—"North African Egg".

The first time I tasted this dish was at the home of Consul General of Israel in Guangzhou—Mr. Alex Goldman. Firstly, he poured olive oil into a large pan and heated it, adding chopped tomatoes, garlic, medium-spicy green peppers, a little bit of chili pepper and tomato sauce. After that, he continued to stir until most of the water in the tomatoes had evaporated. The red tomato sauce looked like a rich, delicious and healthy red curry. Then, he placed two poached eggs onto the tomato sauce, lightly seasoned them with salt, covered the pan with a lid, and simmered the dish for 20 minutes until the yolks were fully cooked. To my surprise, my friend added coriander, commonly used in Cantonese cooking as well, to the freshly cooked Shakshuka. He said he liked coriander very

much, and often used it in cooking other Israeli dishes, such as hummus. In fact, coriander originated from the Mediterranean coast and Central Asia.

In different regions of the world, there are different versions of Shakshuka, ranging from mildly seasoned vegetarian to spicy and salty with proteins. Some recipes add green and red chili peppers for a more spicy variation, while others add feta cheese and sausages or a dash of cumin powder. (cumin powder is as commonly used in Israelis dishes as white pepper powder is in Cantonese dishes.)

The origin of Shakshuka reminds me of my own hometown's Fried Rice, a dish which was also inspired in part by a desire to make the best use of leftover rice, vegetables, eggs, and bacon, and to save time and money.

多吃海鲜菇，为环境多添几分美

根据我的观察，热爱美食的人中，很少有人不喜欢甜食或油脂丰富的美味。这是为什么呢？因为当我们摄入高热量的油脂和糖分时，比如油炸食品或甜点，大脑会分泌一种名为多巴胺的物质。这种物质会刺激我们的食欲，并带来愉悦感。因此，许多人将糖分和油脂视为能激发我们味蕾的两大重要因素。除了这两者外，还有一种来自大自然的味觉享受——鲜味。而这种鲜味最常见于香菇及其他食用菌中。

据日本科学家的研究，食用菌中的鲜味成分主要是鸟苷酸，用它来熬汤，能让高汤鲜美至极；而将食用菌蒸或炒，即便不加额外的调味料，也能带来满满的幸福感和满足感。

目前，在中国市场上常见的食用菌（包括野生和人工栽培的品种）约有240种，其中商业化栽培的有30多种。在我的家乡广东，最受欢迎的食用菌包括香菇、蘑菇、草菇、金针菇、杏鲍菇和海鲜菇。而其中的海鲜菇，是中国南方菇农在从日本引进的蟹味菇的基础上加以改良而培育出的品种。

我曾有幸参加国内最早批量化生产食用菌的践行者之一——江苏润正生物科技有限公司举办的蘑菇宴，并参观了他们位于苏州的食用菌培植基地。席间品尝了鲜美无比的蒜蓉清蒸海鲜菇。这些人工栽培的海鲜菇，生长在一个个装满培养基的塑料袋中，培养基则由木屑、玉米芯、稻草和植物秸秆等废弃物制成。海鲜菇一边"享用"这些废弃物，一边为大自然充当"分解员"。

海鲜菇的长度大约10厘米，色如白玉，顶部顶着一个小冠状的"帽子"，模样清新可爱。我了解到，海鲜菇的最佳烹饪方式就是开边清蒸，这样最能保留其原有的鲜美风味。挑选新鲜、无公害的海鲜菇，清洗干净后用手撕成几片平铺在蒸碟上。准备蒜蓉、葱、油和盐，取一半蒜蓉入油锅爆香，与另一半蒜蓉混合，加入适量的油和盐搅拌均匀，铺在海鲜菇上，入蒸锅蒸3分钟，最后撒上葱花即可食用。入口的那一瞬，鲜美四溢，海鲜菇滑嫩和微脆的口感让人回味无穷。

虽然海鲜菇最初源自日本的蟹味菇，但如今中国的栽培技术已十分成熟，国内的菇菌栽培技术也在迅速发展。中国已成为全球最大的食用菌生产国。

而我，也由衷地爱上了各类菌菇。清蒸或清炒的简单做法能最大限度地保留食材的原始鲜味，烹饪时间也极短。只需几滴油，配以简单的调料，稍加蒸制或翻炒即可上桌。如今，菌菇已成为我每周餐桌上必不可少的美味佳肴。

清蒸海鲜菇
Steamed Seafood Mushrooms

Mushrooms and Fungi—Let's Eat More of Nature's Recyclers

According to my observation, many people love sweet or oily foods. Why is this? From a biological perspective, when the human body ingests high-calorie foods with high fat and sugar content, such as many fried foods or desserts, dopamine is produced in the brain. Dopamine is a substance that stimulates our appetite and triggers happiness. For many people, sweetness and fats are two key components that make foods irresistibly delicious. However there is a third natural food element that can make us "happy":umami. This delicious flavor is often found in the mushrooms and other edible fungi.

According to research by Japanese scientists, the umami in mushrooms that makes the foods so irresistible primarily comes from guanylic acid, a compound that, when used in soups, can add delicacy to the broth. Edible mushrooms steamed or fried can also bring us joy and satisfaction without the need for additional seasoning.

At present, there are about 240 types of edible mushrooms commonly found in the Chinese market (including wild and cultivated varieties), and more than 30 types are commercially viable. In my home province of Guangdong, the most popular edible mushrooms include shiitake, button mushrooms, straw mushrooms, needle mushrooms, king oyster mushrooms, and seafood mushrooms. The "seafood mushroom" was developed by mushroom farmers in Southern China from a crab-flavored mushroom variety originally introduced from Japan.

Previously, I had the opportunity to attend a fungus banquet hosted by Runzheng Biotechnology Company, one of the pioneers of the industrialized production of edible mushrooms in China. I also visited their edible mushroom cultivation base in Suzhou, where I tasted the fresh and rich garlic-steamed seafood mushrooms, full of umami. Human cultivated seafood mushrooms are grown in plastic bags filled with cultured fertilizers, which are made from discarded leftover materials such as wood chips, corn cobs, straw, and plant citrus stems. So these mushrooms are actually helping to decompose waste in nature, making them nature's "decomposers".

A seafood mushroom is fresh and cute, about 10 centimeters long, with the color of white jade, and a crown-like cap. I learn that the best way to preserve their natural flavor is to steam them. Start by choosing fresh, organic seafood mushrooms. Wash them thoroughly and tear them by hand into smaller pieces and lay them flat on a steaming plate. Prepare minced garlic, scallions, oil, and salt. Saute half of the garlic in oil until fragrant, then mix it with the rest of the garlic, oil and salt, stir, and spread evenly on top of the washed seafood mushrooms. Steam for 3 minutes, sprinkle with chopped scallions, and the dish is ready. With the first bite, the rich umami flavor fills your mouth, and the mushrooms' tender yet slightly crisp texture leaves a lasting impression.

Although seafood mushrooms originated from Japan's beech mushrooms, China's cultivation techniques have since matured. China is now the world's largest producer of edible mushrooms, with mature cultivation techniques and rapidly developing technology. As mentioned earlier, the seafood mushroom was created in China by experimenting with a variety of mushrooms introduced from Japan.

As for me, I have truly fallen in love with eating various types of mushrooms. I find that steaming or stir-frying them is the simplest and most effective way to bring out their natural flavors, while also saving time. A few drops of oil, combined with some basic seasonings, little oil, with either steaming or a quick stir-fry is all it takes to create a delicious dish. These days, mushrooms have become a staple on my weekly menu.

最简单易做的法式甜点

一个已经发硬的面包，只需与鸡蛋、牛奶一起烹饪就能做出来一碟美味甜点，这道甜点的法文名字叫作Pain Perdu，中文名字叫作"法兰西多士"，亦是港式西多士的前身。Pain Perdu的直译表述是"丢失了的面包"，指的是那些直接吃颇有点难以下咽、没用了的面包。Pain Perdu历史悠久，相传在中世纪时期的法国，当地经济拮据的家庭常常舍不得丢弃吃剩下的面包，于是想出来一个法子：将剩下的面包切片，蘸上鸡蛋液以及牛奶使其变软，然后用黄油煎成金灿灿的来吃。这个做法很快在当地流传。面包对法国人民来说是如此重要。据说，在18世纪末的法国大革命期间，法国人平均每天能吃掉3～4磅（1.36～1.81千克）的面包，如果出现面包供应短缺或者质量不佳的情况，还会引发骚乱。虽说这道甜品最早是法国当地穷人发明的，但如今喜欢上它的可不仅限于那些经济拮据的人，富人也爱吃。据说，法兰西多士因为在16世纪的时候深受法国国王亨利四世的喜爱而在富人圈子流行，并逐渐流传到世界各地。现今在全球各地的高级餐厅、普通饭馆都能见到其身影，并且衍生出不同的模样。

在这里，我与大家分享的是法兰西多士最传统的做法。准备一碗全脂牛奶，往里面撒少许盐，搅拌开来，再将两个鸡蛋打入另一个碗里，搅拌均匀。面包建议用硬一点的，或者有条件的话，买一个法棍面包，切下不薄不厚的片。平底锅里放入一块黄油，等煎到开始嗞嗞冒油的时候，夹起一片面包放入全脂牛奶中，蘸上牛奶后快速拿起，再放入盛有鸡蛋液的碗里均匀地蘸上鸡蛋液，放入锅里煎。当面包一面煎得差不多呈金黄色的时候，翻过来再把另一面煎至同色。把面包夹起来盛在盘子里，撒上白砂糖，最后根据个人喜好放一点草莓或者蓝莓在上面。如果想吃得健康，可以淋上一点蜂蜜。

如果说这款甜点的主要构成元素是面包，那么覆盖在它上面的元素从来不乏新意：白砂糖、蜂蜜、枫糖浆、咸黄油焦糖、果酱、奶油冻，甚至是香草冰激凌，应有尽有。法兰西多士还可以被做成咸味的——把白砂糖去掉，取而代之的是培根、香肠、奶酪、鼠尾草或百里香。

最近，我在为《Lingling探世界》全球美食节目采访法国驻穗总领事福希玮而做准备，节目里所选择的菜肴必须是有助于环境发展、减少碳排放量的菜肴，我们当时就为选什么法国菜而进行"头脑风暴"，首先想到的就是法兰西多士。对于平日里工作忙碌的我而言，它已成为我周六开启愉快周末的首选早午餐。

已经发硬的面包
Hardened Leftover Bread

法兰西多士
Pain Perdu (French Toast)

Pain Perdu—The Easiest French Dessert You Can Make at Home

In France, there exists a magical dish that transforms a piece of hardened bread into a delightful delicacy simply by combining it with eggs and milk. The French name is Pain Perdu; it is known in English as "French toast", and it is the predecessor of Hong Kong style toast. Pain Perdu has a long history. The direct translation of Pain Perdu is "lost bread", referring to bread that is a bit difficult to eat plain. It is said that in medieval France, economically disadvantaged families had no choice but to be thrifty and couldn't afford to throw away any bread. So they came up with a way to slice leftover bread, dip it in an egg and milk mixture to make it soft, and then fry it in butter. This practice quickly spread locally. In France, bread is very important to the local people. It is said that during the French Revolution at the end of the 18th century, the average French ate an average of 3-4 pounds(1.36-1.81 kilograms) of bread per day. It was such an important food staple,

that a shortage of the bread supply could cause riots. Although this dish was first invented by the working class in France, it soon graced the tables of the rich. It is recorded that in the 16th century, Pain Perdu was deeply loved by King Henry Ⅳ of France and quickly became popular internationally in wealthy circles, spreading across the world. Today, it can be seen in high-end restaurants and ordinary restaurants alike, all around the world, modified for regional palates.

What I am sharing with you here is a recipe for the earliest and most traditional version of Pain Perdu. Prepare a bowl of whole milk, mix in a bit of salt, and stir well. In another bowl beat two eggs. It is recommended to use harder bread, or if conditions permit, a French baguette. Cut a few slices off of the bread and dip each piece of bread into whole milk. Then quickly pick it up and place it into the bowl with the egg mixture. Dip it evenly in the egg mixture and then transfer it into a frying pan and fry in butter. When one side is almost golden, flip it over and fry the other side. Plate the bread, and finish with your of sweet or savory toppings.

In fact, there are a wide variety of toppings which have become popular with Pain Perdu, which can vary from region to region and country to country. Some of the most popular are white sugar, honey, maple syrup, buttery caramel, jam, jelly, whipped cream and even vanilla ice cream. Pain Perdu can also be made savory—just replace the sweet toppings with bacon, sausage, cheese, sage, or thyme.

Recently, I was preparing for an interview with Sylvain Fourriere, the French Consul General in Guangzhou for the *See the World with Lingling* TV program. Dishes selected for the program must be those that are sustainable and reduce carbon emissions. As we began brainstorming French dishes the Consul General's first thought was Pain Perdu. As for me, it has now become one of my first choices for a Saturday brunch to start a happy weekend!